Docks
and
Destinations

Coastal Marinas and Moorage

A User-Friendly© West Coast Cruising Dock-to-Dock
Destinations Guide
San Juan Islands to Ketchikan
Docks • Facilities • Services
• Launch Ramps

Peter Vassilopoulos

By The Same Author
Anchorages and Marine Parks*
Antiques Afloat
from the Golden Age of Boating in British Columbia
**A West Coast Best Seller*

Seagraphic Publications • Vancouver Canada

Photographs, text, drawings and diagrams copyright 1994/1998 Peter Vassilopoulos.
Seagraphic Publications Ltd. Vancouver, Canada. Box 1312, Delta, BC. V4M 3Y8.
In the USA: Box 984, Point Roberts, WA. 98281-0984
Prepress graphics and typesetting Seagraphic Publications Ltd.
Photo scanning and production assistance–Stephanie Bold.
Printed in Canada.
Photographs by author unless otherwise indicated.

The information in this book is accurate and correct as far as can be determined. All cautions and docking information is provided without guarantee and it is up to the boat operator to ensure the proper use of navigational charts and other aids to navigation. Depths are approximate where recorded and should be verified by use of charts, depth sounders, BC Sailing Directions, Small Craft Guide and other sources. Maps and diagrams are approximate and not to scale and should not be used for navigation purposes. Hydrographic Charts should be used at all times when navigating waterways, bays, coves, harbours and marinas. The publisher and author is not liable for marine operations leading to accident, damage or injury in any way connected with reference to this guide. It is intended purely as a reference to available facilities at marinas on the coast.

First Printing–August 1994
Second Printing–March 1995
Third Printing–March 1996. (Revised)
Fourth Printing–October 1998. (Revised)

Library of Congress (Registered)
Canadian Cataloguing in Publication Data:

Vassilopoulos, Peter, 1940–
Docks and Destinations
ISBN 0-919317-26-X 3. 3rd revised and enlarged edition

1. Marinas–British Columbia–Pacific Coast–Guidebooks. 2. Marinas–Washington (State)–San Juan Islands–Guidebooks. 3. Pacific Coast (B.C.)–Guidebooks. 4. San Juan Islands (Wash.)–Guidebooks.
I. Title.
FC3845. P2A3 1998 797.1'09711'1 C98-910907-1
F1089.P2V37 1998

Copies available from marine stores, marinas and book stores. Distribution and acquisition enquiries to Seagraphic Publications (604) 274-4333. Fax (604) 274-4366
email: *divermag@axionet.com*

Peter Vassilopoulos

Docks and Destinations

A coastal guide to marinas and moorage facilities on the West Coast

Featuring the San Juan Islands, the Gulf Islands, Desolation Sound and North of Cape Caution to Ketchikan, Alaska.
Plus marinas on the west coast of Vancouver Island

Includes the San Juan Islands, the Strait of Georgia, the Lower Mainland, Howe Sound, the Sunshine Coast, Desolation Sound, Johnstone Strait, Knight Inlet, Queen Charlotte Strait, Fife Sound, Alert Bay area, Rivers Inlet to Prince Rupert and Vancouver Island Quatsino Sound to Sooke.

Cover: A tranquil early morning at a boating resort dock in Pender Harbour. The signs say it all.

PREFACE

Since the introduction in 1994 of the first edition of ***Docks and Destinations*** I have had the opportunity to revisit many parts of the coast, rediscovering some secluded anchorages and making new friends at marinas and in the boating community along the way. It appears those using this guide have found it useful, as intended, for providing quick reference to facilities and information on places they had not known previously. I am gratified my work has been of value and that I have been able to share some knowledge with those who are out and about in their boats. I encourage new boat owners to enjoy our local waters and to do so in safety.

This guide is a listing of primary ports (marinas) with reference to secondary ports (public docks). It covers docks with and without services. Use it whenever approaching any of its ports in order to know what dock layout and services are available. Mariners have found it particularly useful for telephoning ahead for reservations. *Please refer to* ***Anchorages and Marine Parks*** *(Seagraphic Publications 1998) for comprehensive coverage of the anchorages and marine parks on the BC coast and in the San Juan Islands.*

Boat owners using this guide are kindly asked to observe proper etiquette on the water and at the various marinas and other facilities. At some places there are severe water shortages and mariners are requested to use available water with discretion. Garbage cannot be disposed of easily at most island locations and mariners are asked to not leave their garbage at the docks. Expectations by mariners of the marina operators can sometimes be unreasonable. Please consider the difficulties under which people on the coast have to function. Their season is extremely short—about two and a half months to possibly three months of the summer holidays. They have to make a living and cover their annual costs in those short months. They have to bring in supplies from varying and often long distances. This means not only the produce and groceries which they offer for sale but also their materials to build and maintain their properties. And remember they have established themselves in an area of wet wintery conditions where the slightest neglect results in a quick return of their property back to nature. There are many competent people working at marinas but staff who assist for summer are not always experienced and it is difficult for them to know the specific preferences of individual arriving boat operators. The facilities found along the coast are usually the mom and pop type of businesses and mariners arriving on their doorsteps in need of replenishment are more than welcome, giving the operators an opportunity to meet people and communicate on a friendly, personal level.

You are on holiday. Relax. Be patient. Give them a friendly greeting and you will be gratified with their welcoming response. Operators who welcome you with a grunt rather than a warm, friendly smile, are rare on the coast. If you do come across any such operators, try to be understanding (most people have their bad days once in a while), or just respond by trying an alternative facility next time.

Boating friends and acquaintances have offered advice on what to include and change in this edition. I thank them for their input and have used some of their suggestions. While the final format is mine alone, the input from boating friends has definitely influenced it. I must thank the following for their help:

My wife Carla, Walter and Rita Lee, Henry and Jeanne Karcz, Chris and Sue Fraser and those fellow boat owners who have told me or reminded me about places I may otherwise have missed. Thanks also to Pacific *Yachting Magazine, Boat Journal, Nor'Westing* and others which have had kind reviews on the previous editions. Thanks to the friendly neighbours at marinas we have used as home base

Pulling out of an overnight anchorage at Prevost Island. Many vessels use Prevost as a sheltered stop before continuing to Ganges where there are marinas and lots to do ashore.

and destinations. Their input and encouragement helped push the book and now this latest edition to conclusion. Thanks to the many readers of the previous editions who reported back to me with information on changes and variations at coastal facilities.

In producing this new edition of Docks and Destinations it is my sincerest hope that you will use it to find convenient and safe overnight moorage.

New material for this edition consists mostly of aerial photographs for which I must express appreciation to Heinz Bold and Henry Karcz. I believe that these, presented as large as the format of this book will allow, combined with the diagrams and text, will provide the mariner with a quick reference to the various facilities and destinations that I have included. It continues to provide functional, user-friendly, easy access to information needed and guidance on available facilities. It is therefore presented in its existing format showing principal marinas throughout the coast along with reference and charted location of public docks. Like its companion guide, *Anchorages and Marine Parks*, it is designed to provide guidance but not to remove the joy of exploring for yourself. And it is intended to encourage you to moor at a marina and go exploring ashore where walking the local roads and trails provides good exercise. This new edition includes a listing of marinas on the west coast of Vancouver Island. Thanks to Stephanie Bold for scanning photographs and to my wife Carla and Catherine Adler for assisting me contact marinas by phone and fax for verification of the new information gathered for this edition.

I stated previously that I didn't believe Shakespeare with his copious collection of adjectives and flamboyant superlatives could have found adequate description for some of our landscapes, waterfalls, mountain peaks and deep waterways. In retrospect, he probably could have, but not I. For me it is the use of photographs and other illustrations that help reveal the charm of the islands and inside passages of the area known as the Pacific Northwest. The best descriptions have been coined in the names of some coastal places such as God's Pocket, Village Island, Minstrel Island, Bones Bay, Telegraph Cove, Kingcome, Ocean Falls and many others.

Marinas–Guest Moorage

San Juan Islands and B.C. Coast

Home Marinas

From Sailpast to Summer

Many home marinas are exclusively for privately owned pleasure boats. Here they are stored and maintained throughout the year. As the dawn breaks each year on a new spring and the chill of winter is diminished owners and yacht club members in particular take to scrubbing and polishing their boats in preparation for a colourful sailpast followed by as many boating weekends and prolonged periods away as possible. Cruising to general destinations such as the the San Juans, the Gulf Islands or Desolation Sound is the trend. Frequently there are no specific plans for overnight moorage other than a vague intention to stop if there is suitable anchorage or moorage at one of several possible overnight shelters. Increasingly, and with more boats converging on popular destinations, it is becoming essential that reservations for moorage be made in advance. This guide is intended to help mariners decide where to stop in safe, sheltered moorage overnight and where services, needed by the boat owners or their crews, can be readily acquired. To this end it provides phone numbers, details of marina facilities and other pertinent information.

Heading Out

> Compromises are sometimes necessary when boating and frequently these begin right at your home marina. Consider all the options when planning to leave the dock, monitor the weather and don't be reluctant to postpone a departure for the sake of safety. Home ports or marinas are basically storage areas for boats, some of which seldom are used for long voyages. Most marinas in this guide cater to those who cruise away from home, particularly in summer. Sometimes conditions are not favourable for leaving one's home port as planned and it may be surprising to find there are nearby alternatives, some of which will show up in the following pages.

Most mariners travelling to popular cruising destinations begin their journey from one or another high density population area of Vancouver Island or the coastal lower mainland of British Columbia or Washington State. If you keep a boat in a marina in any of these areas you are probably faced with several possible variations of compromise in your place of moorage as well as your access to holiday moorage and anchorages on the coast. These compromises range from limitations to enjoy your boat at its permanent mooring to your ability to leave or return to the general environs of your marina. In some cases your marina's geographical location is the compromise, or the effects of current and tide where you are moored, or sometimes it's simply the weather combined with an inability to determine adequately what sea conditions prevail until you are well on your way and feel committed to your continued passage.

Victoria and vicinity mariners require a passage around an often rough and tide-ripped Trial Islands and up through a sometimes testy Haro Strait before they reach the placid waters off Sidney. Fortunately most Victoria residents can take one look out to sea and determine the ease of passage. Checking with the tide and current tables is always sound logic to ensure a comfortable beginning of a cruise.

Sidney and Saanich Inlet mariners are in much the same position, but they have the advantage of being where they are going without even leaving the dock (that is—in the heart of the Gulf Islands). And the same applies to Maple Bay, Ladysmith and Nanaimo. Mariners in these areas have enviably easy access for extended seasonal periods to the anchorages and marinas of the Gulf Islands but often look farther afield for their major trips. Their favoured distant destinations include Desolation Sound and beyond. For them access out of their marinas is subject to little more than a decision to go. Swift currents and windy conditions through some passages could delay the more cautious and slower travelling mariner leaving from any of these latter destinations, but usually little compromises their departure and returning plans. At Nanaimo mariners, locals and visitors alike, may await the slack at Dodd Narrows before venturing south into the islands, or may stay at the docks while seas off Entrance Island settle after a storm before crossing the Strait or en route north.

Vessels at marinas on the Sunshine Coast or at places north of Nanaimo are already part way to cruising in Desolation Sound or beyond. For some of them the Gulf Islands are quick and easy to access. But many in places such as Pender Harbour have their boats at that location because of the good fishing and, especially smaller boats, seldom leave the area. Be sure to check your marine charts for area WG and ensure it is safe and that no military exercises are in progress before crossing this part of the Strait.

Mariners on the Vancouver side of the Strait of Georgia are probably the most compromised in their endeavours to leave their moorings and return home. Cautious boat operators will spend days monitoring weather forecasts prior to a major trip, and certainly will listen to the reports on VHF prior to any other departure. Wind and wave height is of utmost interest, tidal changes and currents can be critical and even openings of fishing to the commercial industry can affect one's plans to set off on a voyage. Vessels leaving Vancouver and Port Moody are subject to the currents under Lions Gate bridge and Second Narrows. An easy alternative for weekends out of Port Moody or Vancouver Harbour is Indian Arm, even though sheltered anchorage is limited to Bedwell Bay and little else, and dockside moorage is practically nonexistent, especially for overnight.

Leaving False Creek is quite straight forward and bumpy conditions off Stanley Park are the quick indicator that worse stuff lies ahead, usually at Point Atkinson. Boats departing Richmond and running down the North Arm of the Fraser may reach open water before determining that it was not such a good idea to leave the dock. One can sometimes tell from the wind force at the marina whether there is likely to be rough conditions out of the river, but how rough? Listen to the reports on VHF before leaving. And one can always wait in Coward's Cove for conditions to ease.

Most vessels from these Vancouver and Richmond areas have Howe Sound as their playground. The facilities in Howe Sound are among the best on the coast and increasing numbers of yachtsmen are finding enough satisfaction in spending time at places on Bowen Island or at Gibsons that they decline to cross the Strait to the Gulf Islands other than for extended trips. Out of Surrey or Delta, on the other hand, the Gulf Islands are closer than Howe Sound. And mariners mooring their boats at Ladner, Crescent Beach or Point Roberts can be in the midst of the Gulf Islands in less than an hour (a little longer in a sail boat or displacement trawler). But leaving the Fraser River is one of the biggest challenges on the west coast. (Returning may be another.) The currents at the Sand Heads lightship are vicious at times. A receding tide near low water, especially against a west or northwesterly moderate breeze (don't even think about a strong wind or worse) can be dangerous in the extreme. Refer to the government publication on Weather and Thompson's **Oceanology** for interesting information about current, wind and wave patterns at the river mouth. Bear in mind that while the weather report covers Sand Heads windspeeds it does not provide wave height at the river mouth, a sadly lacking service, especially considering the dangerous nature of the seas at that point. You can always turn back and wait at Steveston for improved conditions or return to your marina, but coming back from across the Strait you reach a point where you are committed.

Leaving Crescent Beach is straight forward enough except that it is a long run across an open bay before entering the Strait. This may be easy and safe when departing, but it is sometimes difficult to assess conditions for returning. Point Roberts is well located for quick, visual assessment of conditions in the Strait and close enough to the Gulf Islands that a crossing of the Strait is quick enough even for slower travelling vessels. But it is very exposed to bad weather conditions for a return trip and mariners should carefully determine what they can expect off Point Roberts before leaving the safety of a comfortable mooring on the other side. One of the compromises for Canadians who keep their boats moored at Point Roberts is the need to clear customs both by road and when cruising. Most who keep their boats at Point Roberts and Blaine feel the compromise is a minor one especially considering benefits of being there, such as an excellent marina with superior facilities and the close proximity to some of the best destinations on the coast, plus the immediate access out of the marina onto the Strait and, for sports fishermen, fishing right off the point. The use of *Canpass* makes border crossing easier. Enquire at Canada Customs for details on *Canpass.*

Boats cruising to Canada out of other Washington ports have a little extra distance to travel through US waters to reach their Canadian destinations. Many simply stay in the San Juan Islands, which have numerous well run marinas and boating facilities. Vessels passing through en route to Canada have to be mindful of wind and tidal conditions at several passages, but with careful VHF weather monitoring and the ability to wait out unfavourable seas, at safe and interesting places, the passage can be most pleasant.

Canadian boats travelling to the San Juans are equally subject to monitoring sea conditions. Know where you are headed and check the route before you leave. Watch for obvious current swept waterways and check the tide and current tables. When cruising any unknown waters check what other boats are doing. If there are no other boats about be doubly cautious and double check your current predictions and weather reports.

Vessels travelling across the border, unless they have a US cruising pass and/or subscribe to *Canpass* must stop at a customs dock for clearance and should carry their clearance reporting number for checking back into their home country.

Once you have managed to clear your marina and the possible obstacles of weather, currents and tide you are on your way to the fabulous Gulf Islands, Desolation Sound or beyond. Make use of the information in this guide and enjoy your coastal cruising.

South to North

The format of this book takes the reader in a south to north progression from one dock to the next. Vancouver Island west coast information is arranged from north to south. The intention is to provide a logical sequence of references to fuel stops or overnight moorage en route to a final destination.

Information accompanying the graphics and photographs is up to date but constant changes are being made at various marinas. To keep your references in this guide updated simply pencil in any new information for your own personal future reference. This book will be revised and updated periodically depending on the frequency and extent to which coastal facilities are altered or improved.

The author welcomes your cruising tips and observations for future editions.

A peaceful early morning in Ganges prior to the busy summer rush in contrast with boats and people at the docks of Telegraph Harbour Marina on Thetis Island in mid season.

Doing it right.

Marinas are a home away from home. When you tie up to a dock at a private marina you are in effect stopping in to visit with other boat owners and with the owners and operators of the marina. What you do and how you operate your vessel says a lot about you and a lot about your experience as a mariner. Your reception from marina managers and fellow boat owners will be determined by the impression you create from the moment you nudge your boat up to the dock.

Some boat owners, and often this applies to novices, don't really care how they are perceived by others. Some want to conform to a certain standard yet will not avail themselves of any form of training or take any advice. Those same people also tend to not learn from their errors. For mariners who wish to fit in, be they newcomers or old hands, here are some basic tips:

During the peak months of summer telephone ahead for reservations. Remember changes are being made constantly and you can expect to find new additions at some docks, name and phone number changes, different regulations, revised services and other variances from the information contained in this guide. Please note the changes for your own future use.

Before you arrive at a marina ensure you are not passing other installations at speed. Slow down well before you reach the dock. Sitting at a dock in West Sound in the San Juans once I saw a large boat came by at full speed leaving a wash in excess of two feet that caused some damage at the dock. He was heading for a club dock at the head of the bay and had totally ignored the existence of the marina tucked in behind the island to his starboard. That same week I saw a similar sized vessel do the same thing entering Bedwell Bay.

Before you enter the marina assess exactly where your assigned slip is. In most marinas you can call on VHF for slip assignment. The dock photographs and diagrams in this book should help you easily locate the slip or general area of the slip to which you have been assigned. As you approach your slip note effects of current and wind and plan your docking manouvres accordingly. It's always easiest to angle in towards the dock against the flow of water or direction of the wind. If you are backing your boat into a tight slip it is even more essential that you are aware of these conditions. Have your crew at the ready and prepared for landing.

Play it the way the pros do. They attach a line to a centre cleat and hold the end coiled in one hand as they step ashore. Snubbing the line to a cleat on the dock when the boat is close to that cleat will keep it there and prevent the bow or stern from breaking away and swinging out, as often occurs when only a line at the bow or the stern is used. Have fenders down at a height compatible with the height of the docks and positioned one ahead of the centre cleat and one near the stern. Once the boat is stopped the skipper can casually step ashore, secure the bow and stern lines and adjust fenders at leisure.

Other people may offer assistance. Usually they expect to be handed a line and all too often when they take it they totally destroy your docking plan. This is usually done by yarding on the bow line, bringing the bow in too close to the dock disabling you from handling the stern. You don't have to pass them a line. Do so if absolutely necessary or when the skipper has the boat docked. No harm saying to a person "here is the line, please just hold it" or have two docksiders haul in bow and stern simultaneously. Don't try to look as though you are proficient if you cannot pull it off. You are better off to say to anyone watching as you are approaching your slip that you are new at this and would appreciate some experienced help in docking. You'll be surprised how readily people will come to your aid and how pleasant they will be when you are up front about your docking abilities.

Above: The partially drying canal between Thetis and Kuper Islands allows passage at high tide between Clam Bay and Telegraph Harbour.

Docking is just the beginning. You will not enamour yourself or your crew to anyone at the dock if you yell, either at them or at your crew. At many marinas there are full-time personnel employed to assist boaters docking. They are not always the most experienced but as long as you follow some of the above advice their assistance will enable you to perform a good landing.

If you are tying up parallel to a long open dock, tie up your boat in such a way that you allow maximum room for the next boat coming in. If you have a dinghy in the water tuck it in under the bow of your boat while you are not using it in order that you leave room for the next boat.

Anchoring off and going ashore to visit a marina and its facilities, perhaps to have a meal, a snack or browse for some souvenir is a common practice during summer. Also common is the individual who goes ashore only to drop off a huge bag of garbage. On small, remote islands, this is a major problem. If you are a mooring guest you may do well to assess how convenient it is to leave garbage behind, even though some places have disposal bins. If you know you are continuing to a mainland facility soon, save the garbage for that stop. The same applies to water. If you are travelling to an island with limited water resources fill up before going there and use water sparingly. Do not use water to wash your boat.

Some boat owners have been known to tie up at a marina for long periods during the day, fill up with water, perhaps use the facilities such as shower and laundry and then take off and anchor across the bay for the night. There are times when such use of moorage has denied a prospective overnight moorage customer space to tie up. Marinas have a very limited season in which to prosper and mariners who cause lost overnight moorage will not be appreciated.

Most marina operators will be happy to have you stop for a short period if you don't plan to spend the night. Some have a charge per hour. Others, like government docks, allow two hours free and then an overnight charge is levied.

The coast and islands of British Columbia and the San Juans

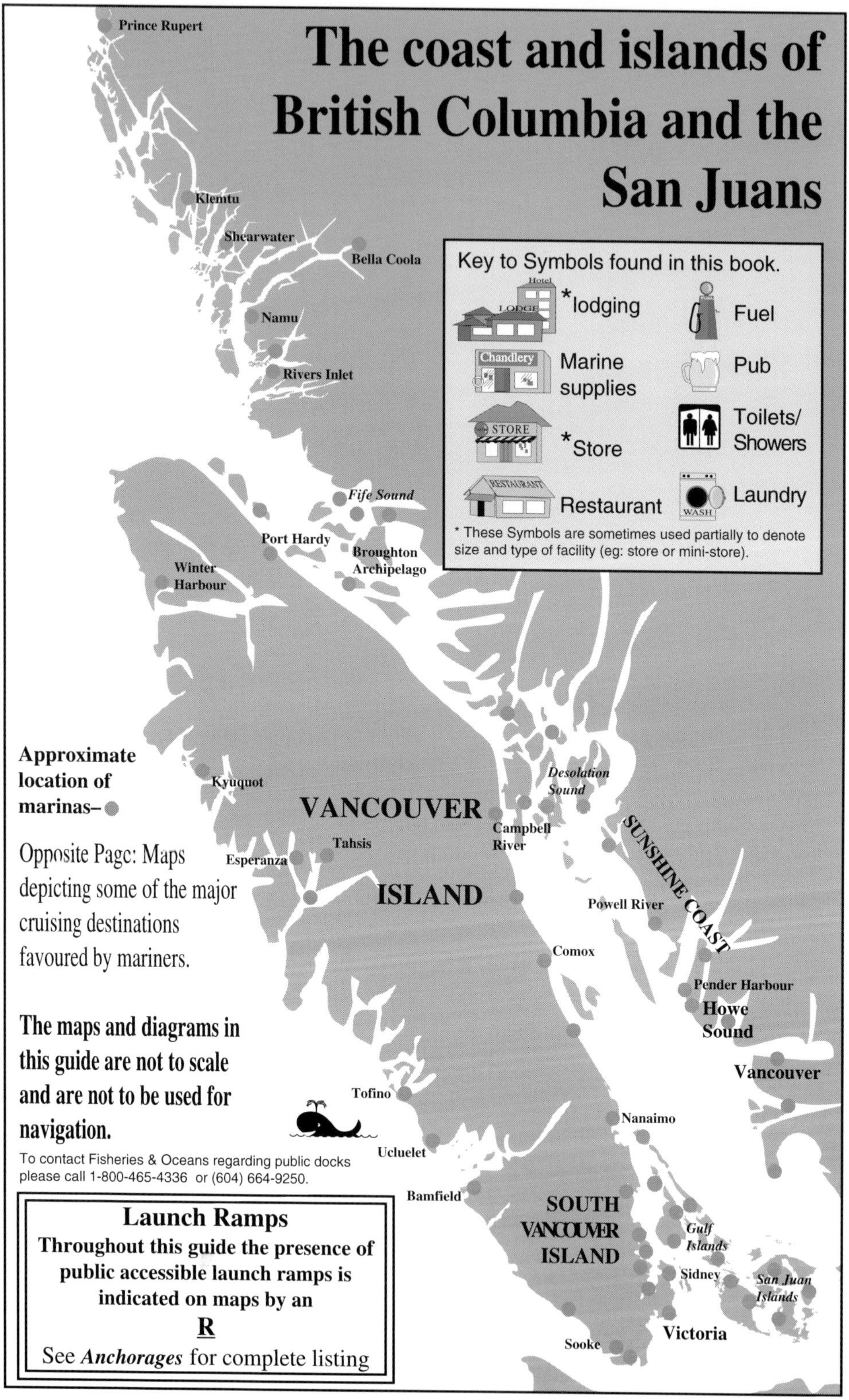

Approximate location of marinas– ●

Opposite Page: Maps depicting some of the major cruising destinations favoured by mariners.

The maps and diagrams in this guide are not to scale and are not to be used for navigation.

To contact Fisheries & Oceans regarding public docks please call 1-800-465-4336 or (604) 664-9250.

Launch Ramps

Throughout this guide the presence of public accessible launch ramps is indicated on maps by an

R

See *Anchorages* for complete listing

Maps and drawings are not for Navigation.

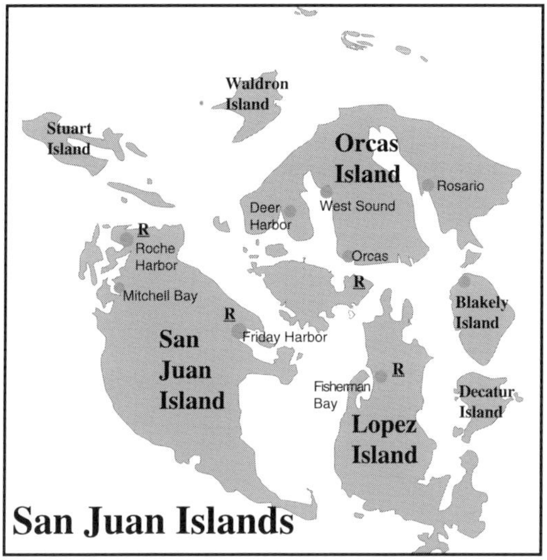

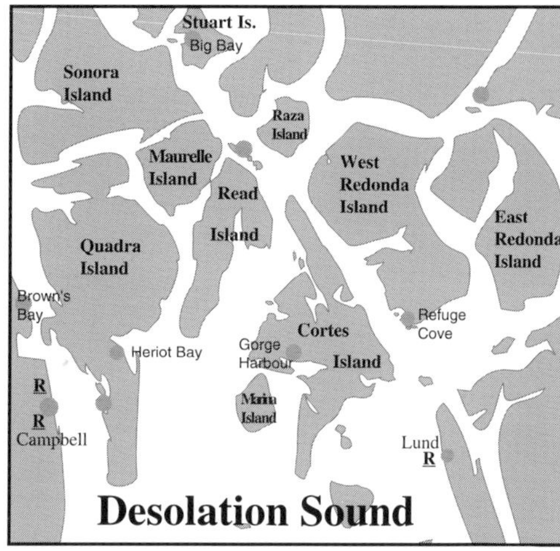

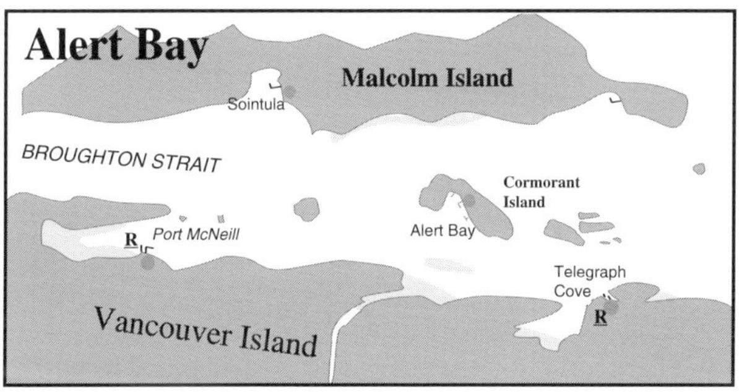

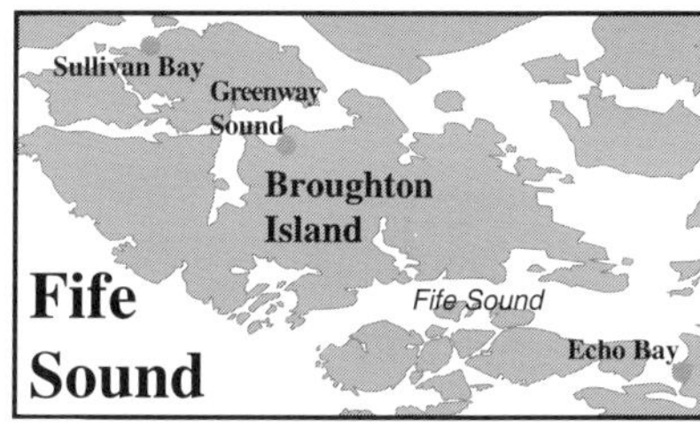

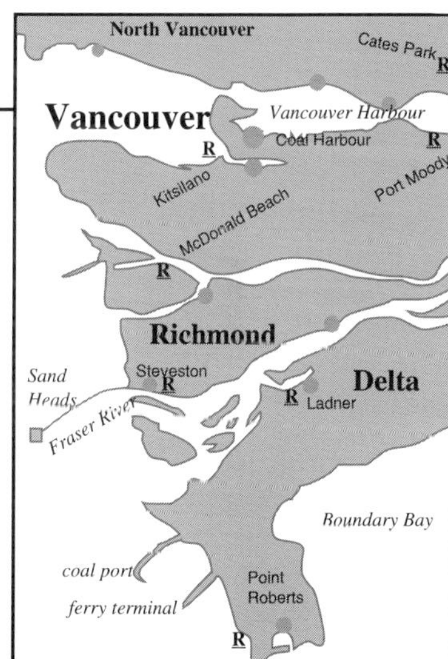

Please use official Hydrographic Nautical Charts when cruising coastal and inland waters.

Preedy Harbour on Thetis Island with a view towards Chemainus on the east coast of Vancouver Island. Many mariners staying at Thetis Island marinas take the ferry from this bay to Chemainus for a day excursion to the shops or to view the famous murals for which the town is known internationally.

Lighthouse at Green Island off the east side of Dundas Island en route Prince Rupert to Ketchikan. A lighthouse is a beacon for all who go to sea, and a welcome sight for mariners in fair or stormy weather.

Docks and Destinations

This book is aimed at providing mariners with guidance to protected overnight moorage at selected public docks.

As you travel up the coast out of your home marina you may ponder your course to your final route. Or you wisely will have planned it well in advance. As you plan your route this guide will help you finalize and consummate that plan by enabling you to call ahead and make reservations one dock ahead for your progress along the way. Your timing for arrivals and departures at your chosen marinas will be subject to the weather and sea conditions and any astute boat operator will make allowances for this as do the marina operators. Although they are sometimes fully booked in advance, no-shows can mean a space for a late arriving boat that has either made no reservations or has been accepted on a stand-by basis. It is wise if you are arriving at a marina on stand-by and the docks are possibly going to be full that you choose in advance a marina where there is an alternative dock or anchorage nearby, preferably one that offers total protection in all weather conditions. Please refer to my companion guide ***Anchorages and Marine Parks.***

The geography of the book accompanies the mariner out of mainland Washington State, Vancouver Island South and BC's lower mainland. It begins with destinations in the San Juan Islands, continues from Victoria, B.C., and on up the east coast of Vancouver Island, taking in Sidney, Saanich Inlet, Cowichan, Maple Bay and on through Crofton to Ladysmith. Then it returns to the Gulf Islands and follows them geographically, more or less up to Nanaimo. It goes across to Greater Vancouver with its Fraser River, False Creek, Burrard Inlet and Indian Arm facilities. Then continues into Howe Sound and on up the Sunshine Coast with its inlets and harbours, until it reaches the entrance to Desolation Sound. It moves across to Comox and Campbell River then on to Desolation Sound and vicinity before continuing up through Big Bay to the world beyond Desolation Sound. Through the rapids and passages it goes up Johnstone Strait, acknowledging Kelsey Bay and branching off to Minstrel Island before continuing to Alert Bay. At this juncture it heads through the Broughton Archipelago to Fife Sound via Echo Bay to Sullivan and across to Port Hardy and then God's Pocket in the Bull Harbour area. From there it takes a route through the Discovery Coast on to Ketchikan, Alaska and ends with a listing of marinas on the west coast of Vancouver Island.

All of the marinas listed have protected moorage and facilities in varying degrees. Some offer limited casual moorage but no marina on the coast, including those operated for permanent mooring clients only, will not turn away an overnight guest if there is room. And it is only in the months of July and August as well as some weekends and long weekends in other boating months that space at marinas is at a premium. Many mariners who operate their boats year round enjoy the quiet and solitude of marinas during the off season. Monitoring weather conditions allows access from home ports to the destination of choice which is usually one that assures complete protection against changing weather conditions.

Historic buildings and places of interest can be found adjacent to many waterfront landings such as the Orcas Hotel, top, and the West Sound Store, above.

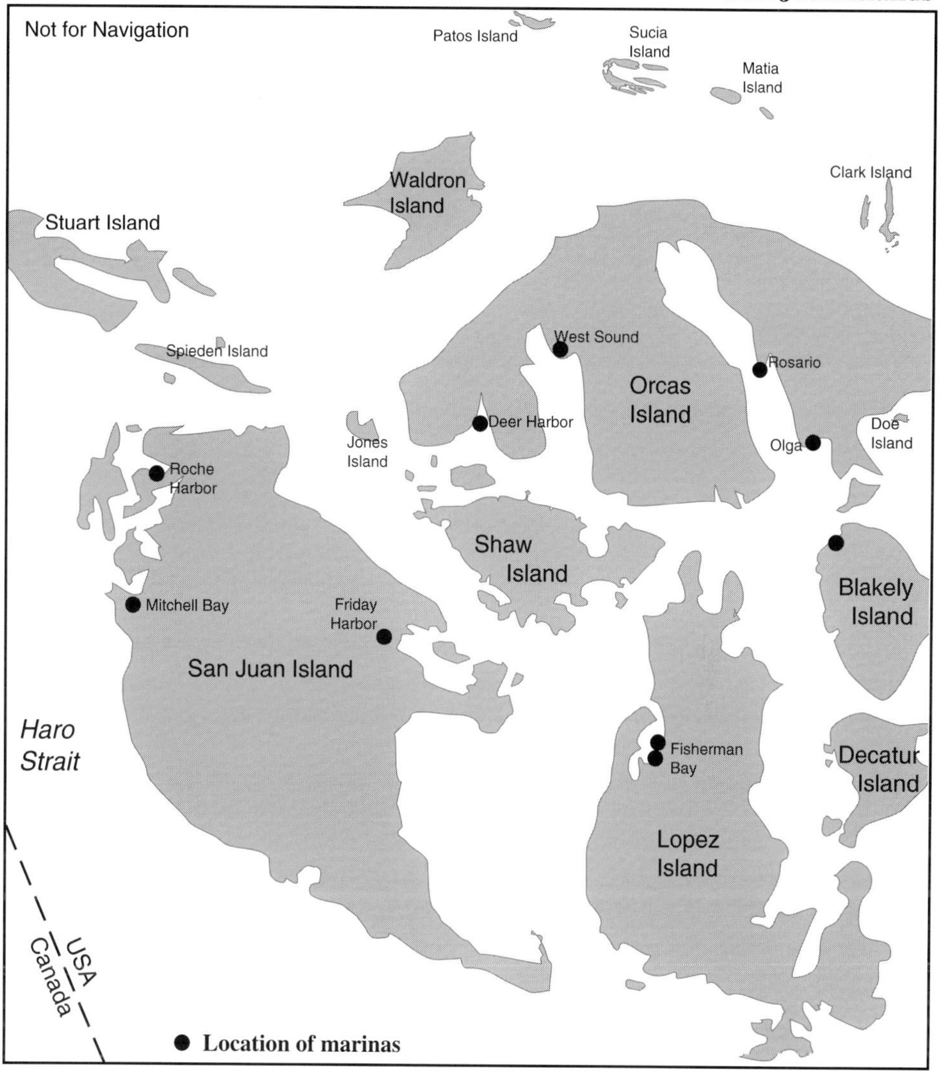

San Juan Islands

Journeying into the San Juans mariners have the pick of some outstanding marinas, fine restaurants, well-stocked marine, grocery and hardware stores as well as arts and crafts centres. There are fast food places, ice cream vendors, pubs, hotels, good accommodations and rustic bed and breakfast places. There are walking and hiking trails and roads as well as car, bicycle and motor scooter rentals. Some stops include dining at places such as Sean Paul's restaurant at Deer Harbor or the quaint West Sound Store and Deli for a good bowl of soup, a sandwich, salad or dessert. At Friday Harbour there is such a wide choice of

restaurants as well as other stores it would take an entire vacation to enjoy the place to its fullest. The annual Jazz Festival on the last weekend in July is a busy time so get in early or keep away if you want to avoid crowds. A more tranquil place may be Fisherman Bay with its good anchorage and marinas, nearby Lopez Village and pleasant dining facilities. Roche Harbor is a busy customs port and attracts some of the larger cruising yachts to its busy marina. It's a place to visit if you like to look longingly at some of those mega vessels that frequent and even monopolize it. The Hotel De Haro usually has fine dining and it's entertaining to watch the evening color ceremony at sundown. Fascinating history of the islands include the early explorations of the Spanish, the presence of the English and the famous Pig War which nearly led to an international confrontation between the British and Americans. One of the more charming remnants of recent history of the islands is Rosario Resort on Orcas Island. A stop at this facility will provide sheltered moorage as well as a chance to acquaint yourself with its splendid history.

A short or long stay in the San Juans can provide a complete vacation, and many Canadians make the trip once in a while just as their American counterparts are steaming through the San Juans en route to the Canadian Gulf Islands and points beyond.

Above left and right: Friday Harbor. Opposite centre: At Lopez Village. Right: The marina at Roche Harbor with anchorage in Garrison Bay beyond. Left: Rosario Resort on Orcas Island.

A smart crew will not pass dock lines to someone on shore until the skipper is good and ready. Spectators who are eager to assist often are inexperienced and will pull on the bow line, throwing the skipper's manouvres totally out of control. It is best only to pass someone ashore a line connected amidships, and only when it is really necessary, such as when wind or current are pushing the boat away from the dock.

Roche Harbor

Roche Harbor Resort & Marina

Kevin Carlton
Box 4001, Roche Harbor,
Washington. 98250
Phone: (360) 378-2155 Fax (360) 378-9800
Chart 18421/33 VHF call 78
email: *roche@rocheharbor.com*
Web: *www.rocheharbor.com*

Marina services:

Moorage. 377 permanent and transient slips to about 180 feet. **Pumpout** facilities. **Power** at docks: 20 30 50/ 220 amps. **Water. Fuel :** Gas, diesel. Propane. Oil.

Customer services:

Showers, laundry, washrooms
Public phones ashore. Phone hook up.
Walking: Road access walking, cycling.
Adjacent and nearby facilities: Store–groceries, a wide range of provisions. Hotel, fine dining, accommodations. Arts and crafts. Clothing, apparel, gifts, groceries. Ice, fishing tackle, licences, marine supplies. Moped rentals. Horse riding. Hiking trails. Boat rentals, kayaks. Good fishing nearby. Pool and tennis courts. Jazz festival late July–call for information. Fourth of July celebrations and fireworks. Color ceremony each sunset. Coffee wagon.

All facilities open March through October. Moorage and basic facilities open year round–partial during winter. Launch ramp nearby. Taxi or bus service to island centres and ferry to Anacortes. Airfield. Major Customs Port of Entry.

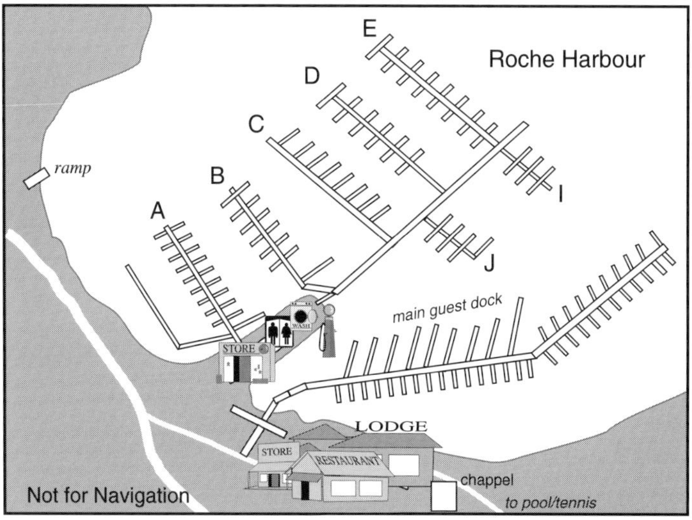

Below: Facilities at Roche Harbour include the well stocked store whose name commemorates the historic Lime Stone company for which Roche Harbour was founded. Opposite, below: Approaching the fuel dock.

Hotel de Haro is the historic landmark at this famous harbor. It is a magnificent building exemplifying the type of construction and opulence of its day. There is a lot to see and do at Roche Harbor. The resort is a museum in its own right, its buildings dating back to 1886. It was founded in conjunction with the adjacent lime kiln and barrel manufacturing company.

Massive changes and additions were made to the marina in 1996/7.

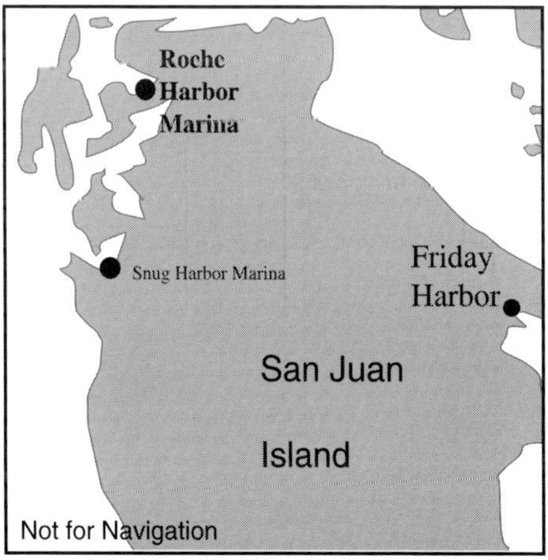

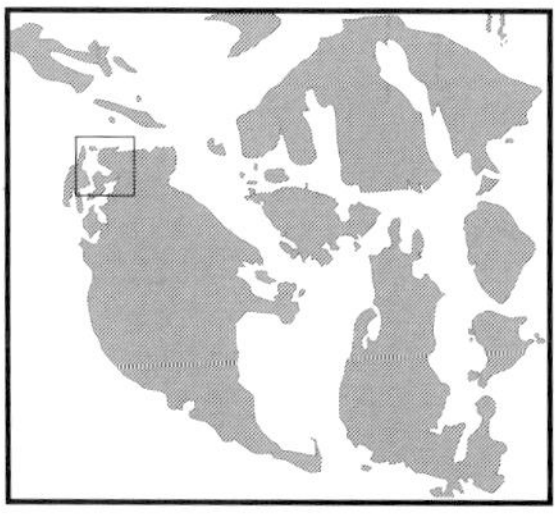

Friday Harbor

Port of Friday Harbor

Box 889, Friday Harbour,
Washington. 98250
Phone: (360) 378-2688
Chart 18434. 18421. Waterproof #43

VHF 66A (US mode)

Marina services:

Moorage. Guest and permanent.
Power at docks: 30 amp.
Water.
Fuel dock adjacent: Gas, diesel, propane.

Customer services:

Pump-out station.
Laundry, showers, ice, bait.
Public phones ashore.
Nearby church/es: multi-denominational.
Pharmacies and other necessities.
Scuba diving arrangements and charters–ask at nearby dive store or marina for details.

Walking: Road access walking, cycling, car and scooter rentals.

Entertainment:

Regular annual music festival late July, Pig War Barbecue in June. Many other events.

Nearby facilities: Ferry to Anacortes. Marine stores–charts, marine hardware, supplies, books, fishing licences, tackle, etc. Post office, liquor, restaurants, banks, accommodations, pubs and specialty stores. Golf, cinema, airport.

Note:
When arriving at Friday Harbor from Canadian waters first check in at the customs dock on the breakwater.
The flag up means the office on the dock is open. Down means you can check in at the office ashore or by telephone if that office is closed. Then proceed to dock A for slip assignment or call on VHF 66A.

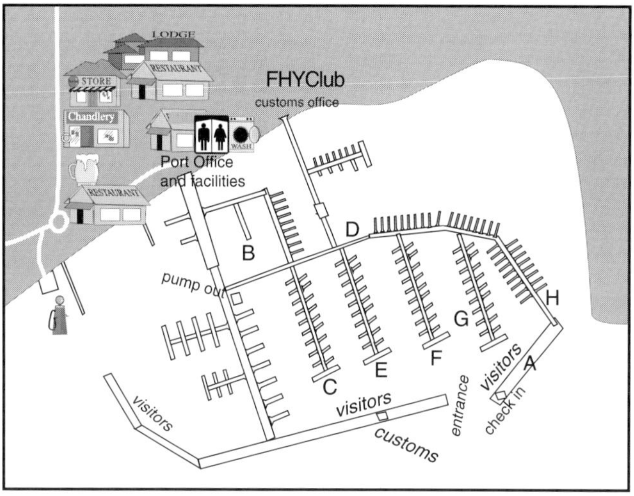

Not for Navigation

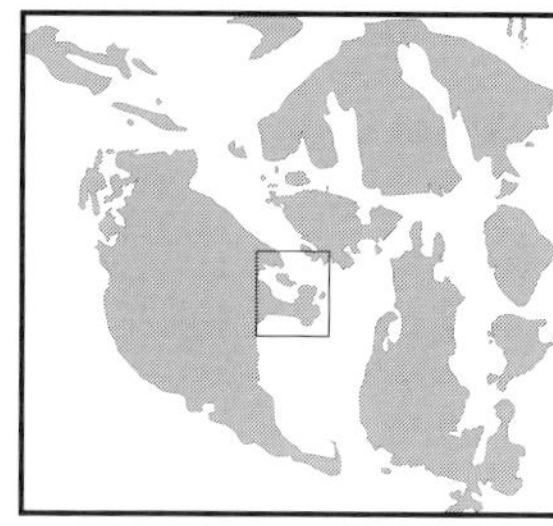

Major Customs Port of Entry:
Phone (360) 378-2080.
Or 1 800-562-5943

Friday Harbor

This large marina is one of the busiest in the Pacific Northwest. It is a major customs stop entering United States waters for Canadian boat operators and one of the major centres for returning American mariners. Located on the east side of San Juan Island it competes with Roche Harbor as a customs stop but the two are vastly different. Friday Harbour has a large town comprising everything from city hall to cinema, supermarkets to specialty shops, hardware store, marine chandleries and a wide variety of restaurants, pubs and bistros. As a major ferry landing, Friday Harbor sees the coming and going of a vast number of people: islanders, cyclists coming to visit, campers, boaters and fly-in sightseers. During summer crowds of boaters flock to the town for the major event of the year, the Summer Jazz Festival. A Pig War barbecue is held in June.

For the boat owner there is fuel, moorage, water, showers, laundry, 30 amp electrical service and all the amenities one could imagine necessary for a major stopover. There are two anchorages in the immediate proximity of the huge marina, one adjacent to the north west marina entrance and the other out in the middle of the harbor.

Snug Harbor

Snug Harbor Marina Resort

Dick Barnes (owner Glenn Kalmus)
2371 Mitchell Bay Rd., San Juan Island, Washington. 98250
Phone: (360) 378-4762 Fax 360 378-8859.
Chart: Waterproof #43. 18421

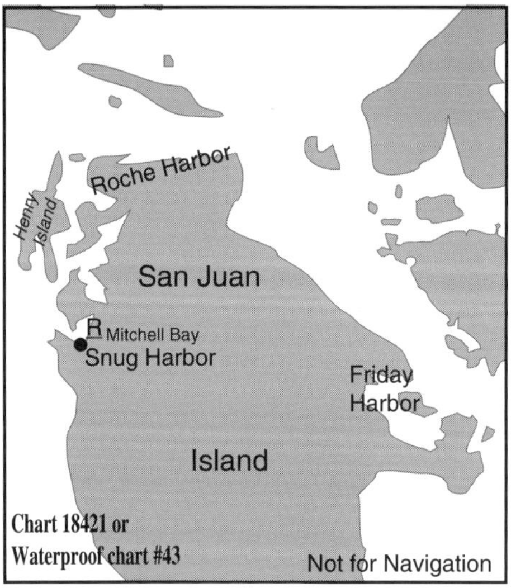

Snug Harbor Marina in Mitchell Bay is protected from the open waters of Haro Strait by a shallow entrance and drying reef. The channel at zero tides drops to three feet. The bay lies south of Roche Harbor.

Marina services:

Moorage. 71 slips. Guests welcome.
Power. 15, 20, 30 amp.
Water.
Fuel : Gas. Repairs. Marine supplies.

Customer services:

Garbage bins.
Shower, laundry. By arrangement.
General store, groceries, provisions, charts, books, clothing, gifts, hardware, marine, fishing equipment. Public phones ashore.
Marina upgrades underway (1998/9)
Walking: Road access walking.
Adjacent Facilities: Accommodations. Ten self-contained, fully equipped bungalows. RV park. Launch ramp. Scuba charters, whale watching.
Note: Rocks in entrance. Channel to left. Depth 3' at zero tides.
Use large scale chart or Waterproof #43 for navigating inside Henry Island.
Nearest customs at Roche Harbor.

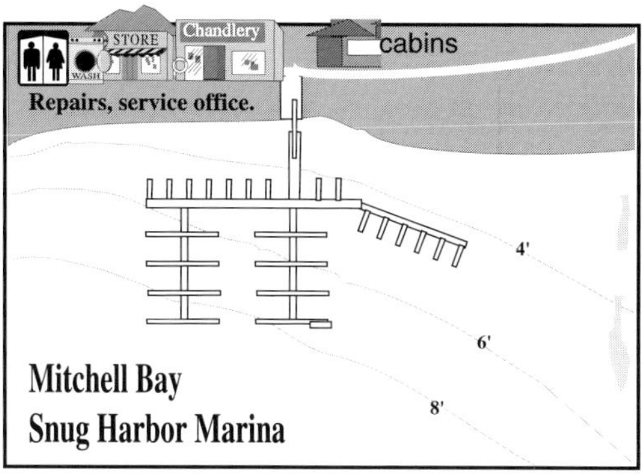

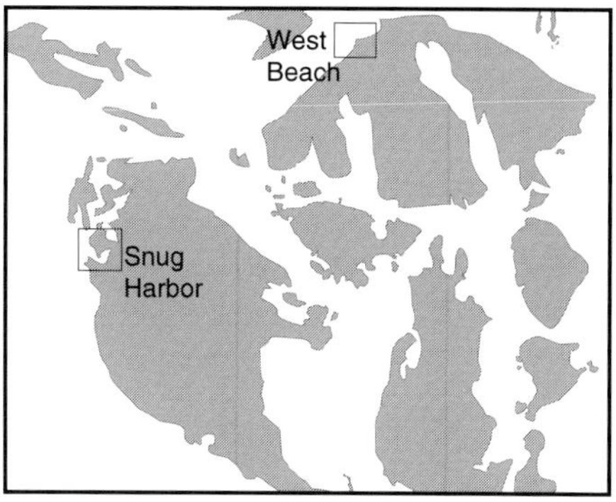

Not for Navigation

Below: West Beach Resort

Orcas Island

West Beach Resort

Tim and Liz Hance
Route 1. Box 510,
Eastsound, Washington. 98245
Phone: (360) 376-2240 fax (360) 376-4746
Chart 18432, 18421
email: *vacation@westbeachresort.com*
Web: *www.westbeachresort.com*

Marina services:

Moorage. Transient boats to 26'.
Water. Fuel: Gas. Propane, ice.

Customer services:

Fish cleaning facility. **Showers, laundry.** Resort. Store, coffee shop. Charts, books, fishing tackle, bait. Stove oil, kerosene.
Entertainment: Walking–road access. Good scuba diving in vicinity.
Adjacent Facilities: Accommodations. 13 self-contained, fully equipped bungalows. RV park. Scuba airfills. Public phone. Launch ramp guests only (fee includes parking for trailers).

Deer Harbor

Deer Harbor Marina

Craig Webster
Box 200, Deer Harbor, Orcas Island
Washington. 98243
Phone: (360) 376-3037 Fax: 376-6091
Charts 18421/34 VHF call 66 (US mode)

Marina services:

Moorage. Transient and permanent.
Power at docks: 15 amp.
Water.
Fuel: Gas, diesel.

Customer services:

Barbecue area. Accommodations.
Showers, washrooms. Ice.
Public phones ashore.
Walking: Road access walking, cycling.

Entertainment:

Small boat rentals, whale watching tours, kayaking, sunset cruises and fishing charters. Heated swiming pool and spas.

Adjacent facilities: Store–groceries, deli and provisions. Store–gifts, postcards, clothing and arts and crafts. Sean Paul's restaurant–elegant dining.
Taxi service to island centres and ferry to Anacortes. Seaplane service to Seatac.

Note:
Clear customs for USA destinations at Roche Harbor or Friday Harbor.

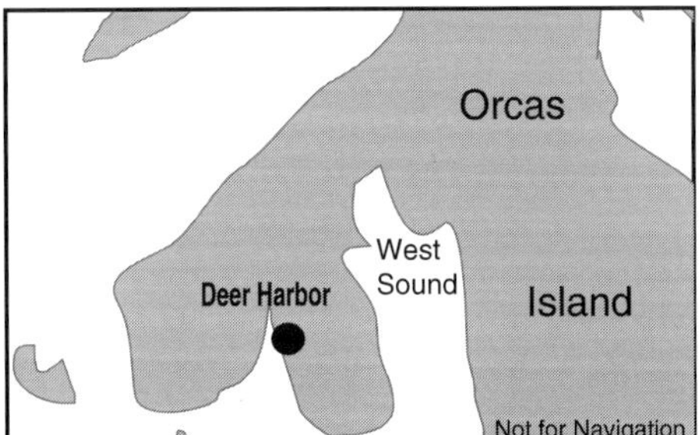

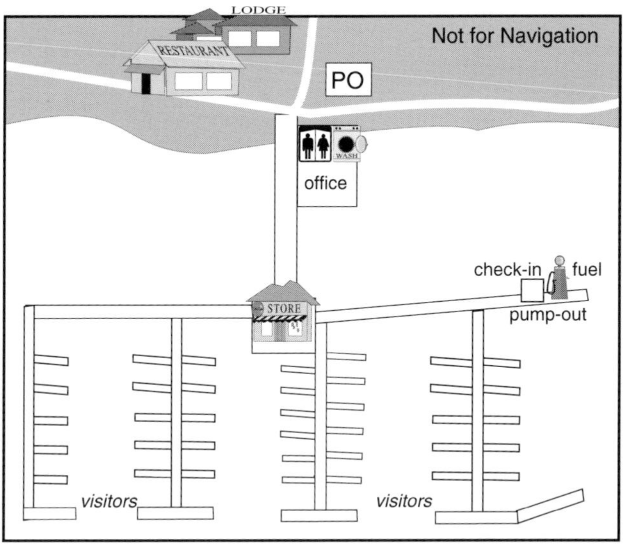

Constant change: Even during revisions of this guide, the layout of the marina was changed– recent photo below. Additional fingers were graphically created in the opposite picture taken during the busy summer season.

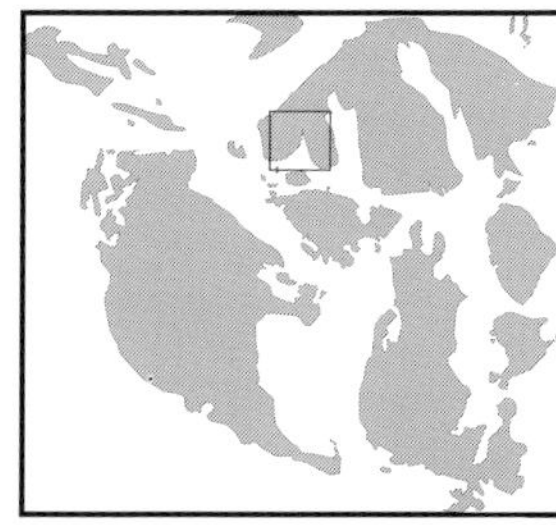

Inset left shows check-in and fuel dock at Deer Harbor. Inset above shows marina with new additional fingers.

Deer Harbor

This cosy corner of Orcas Island is a pleasant stopover at docks that face a semi-protected open bay. Most wind conditions do not bother boats at the marina but a southerly or south westerly wind may cause a bit of movement, especially off season. In season many boats visit the docks to stay overnight and enjoy excellent cuisine at Sean Paul's Restaurant. The marina offers fuel, water, showers, ice and the opportunity to relax at the adjace resort in the comfort of a well maintained swimming pool and hot tubs.

West Sound

West Sound Marina

Mike & Peggy, Ian, Betsy Wareham.
Box 119, Orcas Island
Washington. 98280
Phone: (360) 376-2314
Chart 18421/34 VHF call 16
Hazard: Island reef near guest dock.

Marina services:

Moorage. Some transient moorage.
Power at docks: 15, 30 amp.
Water.
Fuel: Gas, diesel, oils.

Customer services:

Marine ways and repairs. Major shipyard. Complete marine chandlery.
Showers, washrooms. Ice. Public phone.
Walking: Road access walking, cycling.
Nearby facilities: Store–deli and provisions. Light meals. Bed and Breakfast. Taxi service to island centers and ferry to Anacortes. West Sound Yacht Club. Public dock at head of bay.

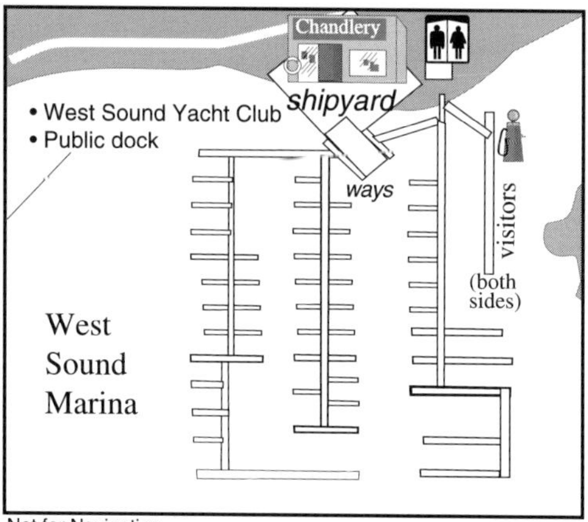

Not for Navigation

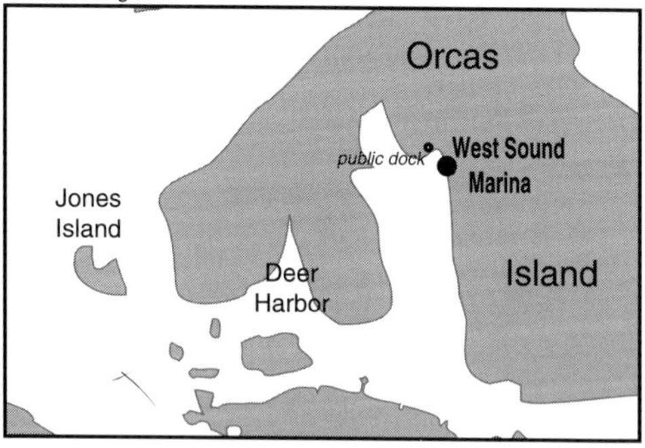

West Sound

This protected moorage is largely for long term resident boats. However, it has an excellent marine service facility with 30 ton travel lift and comprehensive repair services. It has a well stocked marine store and the buildings ashore include a large, modern, heated bay where refinishing and mechanical work can be done in any weather conditions. The facility caters to all of the San Juan Islands with a 24 hour emergency service. Transient moorage accommodates a number of boats on a 250 foot long finger that is an extension of the fuel dock. Gas and diesel are available and are generally priced competitively with fuel at places such as Friday Harbor. It has a pump out station.

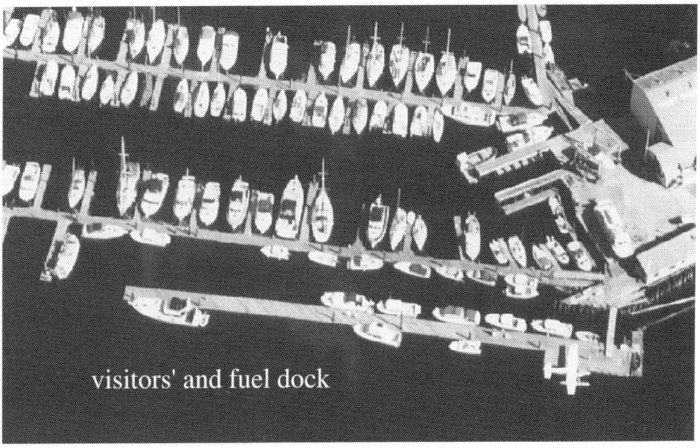

A short walk up the road is a deli which serves hot light meals at lunchtime. An island taxi service out of East Sound is available for transportation to other parts of Orcas Island such as Orcas ferry landing or the town of East Sound. The facility is owned and operated by Mike and Peggy Wareham, son Ian and daughter Betsy. The chandlery offers Johnson outboard products, fishing gear, marine supplies of a wide variety, propane and ice.

Note:
Clear customs for USA destinations at Roche Harbor or Friday Harbor.

Note the docks at the head of the bay. One is the yacht club and the other a public dock.

Orcas Landing

Russel's Store (Independent operator)
Box 200, Deer Harbor, Orcas Island
Washington. 98280
Phone: (360) 376-4389 (Russel's Inc.)
Chart 18430/4, VHF call 66 (US mode)

Fuel dock–Island Petroleum Services

Phone (360) 376-3883.
Moorage. Stopping briefly only.
Water. Fuel: Gas, diesel. Oils.

Customer services:

Stores and facilities ashore. Ice.
Public phones ashore.
Walking: Road access walking, cycling.

Entertainment:

Sightseeing by bicycle popular.
Adjacent Facilities: Stores–groceries, gifts, clothing, arts and crafts. Orcas Hotel nearby. Taxi service to island centres and ferry to Anacortes.

Orcas Landing

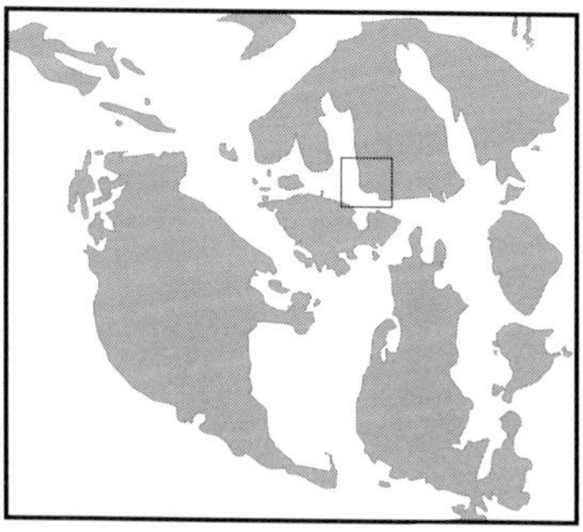

A quaint village at the ferry landing serves the community on Orcas Island. It has an historic hotel as a focal point. The Orcas Island Hotel has rooms and pub and restaurant, the latter of which is reputed to serve excellent meals. Through the week food is served at the pub or at the pub's outdoor terrace area. Changes and improvements should provide greater comfort. The dock at Orcas is for transient mariners

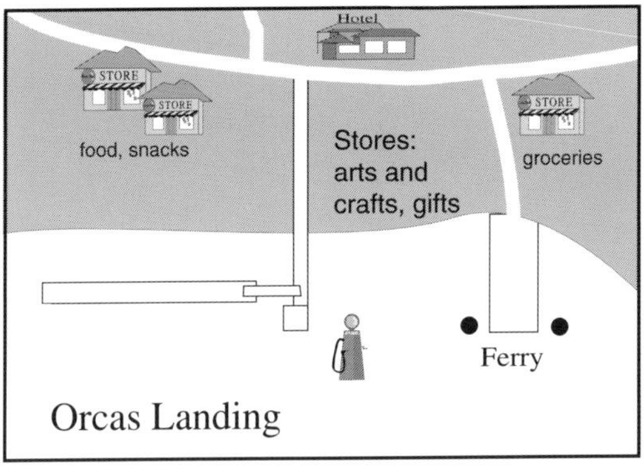

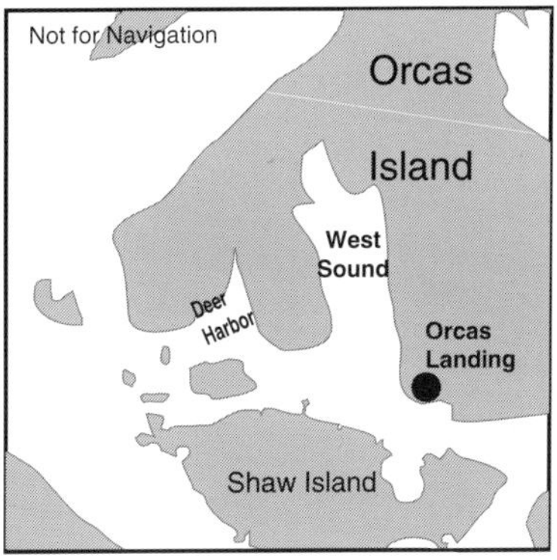

only. It is long and accessible both sides although the shoreward side has a narrow entrance and if a boat is tied up at the end entry for anything but a small boat would be tight. Fuel is available at the dock as well as other marine products and services. On shore there is a gift and craft store, Russels, which has a wide selection of wares. The settlement is popular among transient mariners and land based visitors alike. Summer sees thousands of cyclists, campers and motorists arriving on the island in a constant stream off the ferries. At Orcas it is a common sight for crowds of vacationers to be lining up for their trip home mingling with those just arriving as they flood into hotel, craft stores, grocery store, fast food restaurant, ice cream store or other facilities.

Left: The small craft dock at Orcas Landing.
Below: Orcas Landing with Russel's and other buildings adjacent.
Opposite: The ferry at Orcas Island. The dock for small craft can be seen clearly to port of the vessel.

Rosario Resort Marina

Harbor Master
Orcas Island
East Sound, Washington. 98245
Phone: (360) 376-2222 Fax 376-3680
Chart 18421, 18430/34. WP #43
email: *info@rosarioresort.com*
Web: www.rosarioresort.com

Marina services:

Moorage. Guest and permanent. **Fuel.** **Power** at docks: 30, 20 amp. **Water.**

Customer services:

Laundry, showers, ice, bait.

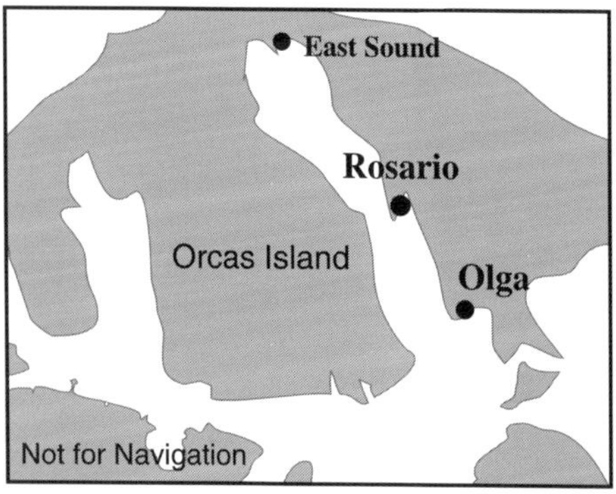

Public phones ashore. Garbage bins. Hotel, pub, spa, pool (indoor and outdoor), retaurants–fine dining or casual. Gift shop. Accommodations and all hotel services. Nearby church.
Scuba diving arrangements and charters–ask at dive store or marina for details.
Walking: Road and trail access walking. Car rentals.

Entertainment:

Regular live music, organ recitals in the historic mansion. Slides and Rosario history narration. Golf nearby. Tennis courts. Kayak tours. Air tours. Whale watching tours. Fishing charters.

Adjacent facilities: Cascade Bay Cafe and grocery store, supplies. Rental cars. Kayak rentals.

Note: Reservations recommended for marina or resort, in peak summer months particularly. Mooring buoys and anchoring landing fee includes passes to spa and pools. **Launch ramp.**

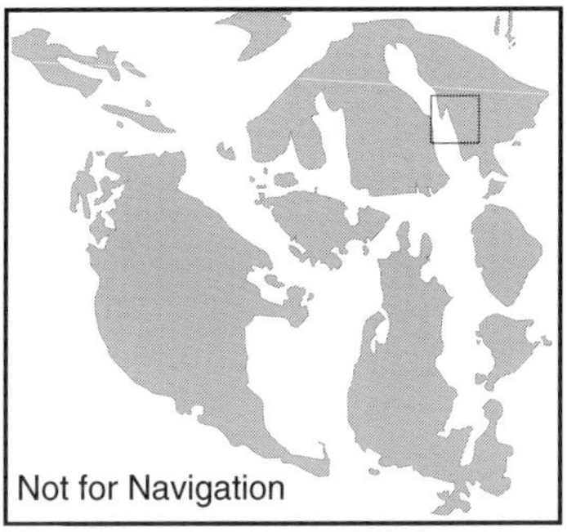

The docks (inset left) are protected behind a breakwater with a fuel dock at the entrance. The mansion, right, looks over the sweep of East Sound. Lower: The landing at Olga.

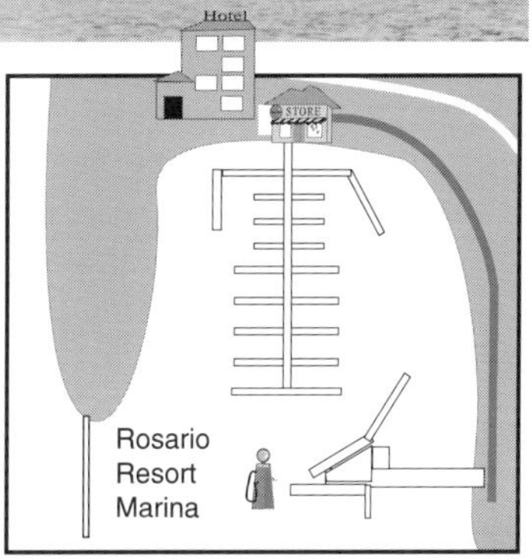

Rosario Resort

There is sheltered moorage at the docks behind the breakwater. Mooring buoys are not necessarily sheltered from the wind and waves that blow up East Sound some afternoons. The hotel is a feature of the San Juan Islands with a colourful history and excellent restaurant. Facilities include water, showers, store and a swimming pool with hot tubs.

Robert Moran, who built Rosario Resort after retiring in 1904, was a former shipbuilder in Seattle. He used shipbuilding methods and materials in the construction of the mansion. The building's walls are made of 12 inch concrete and the roof is sheathed in copper. Windows are 7/8 inch plate glass and many sections of the interior are panelled in mahogany. One of the resort's major features is the Kimball pipe organ, said to be the largest installed in a private residence in the United States. The mansion was listed in the National Register of Historic Places in 1979. The estate was sold in 1938 and the current ownership created the resort in 1960.

Rosario is the closest sheltered marina for accessing East Sound, a bustling village that attracts many visitors annually, mostly by ferry. There are restaurants, stores, banks, post office, medical clinic, movie theater, pharmacy, churches, galleries and many other facilities. East Sound airport provides tours for visitors wanting a lofty look at the islands.

The town of Olga has a quaint village store and community docks. Stop and dine at the village arts co-op restaurant.

A sign on the dock reads: “Dock maintained with community labor and 25 cents a foot moorage fees collected after 6 pm.” It is exposed and not recommended for overnight.

Blakely Island

Blakely Island General Store and Marina

Peter and Robin Lamont
Blakely Island
Washington. 98222
Phone: (360) 375-6121 Fax 375-6141
Chart 18421 18430
email: *blakely@rockisland.com*
Web: *www.rockisland.com/~blakely/*

Marina services:

Moorage. Guest moorage.
Fuel: Gas, diesel.
Power at docks: 30 amp. **Water.**

Customer services:

Showers, laundry, washrooms.
Store. Provisions. Fresh produce, some groceries. Clothing, gifts. Marine hardware. Tackle, bait. Fishing reports taken. Licences. Ice. Hot dogs, snacks. Patio. Banquet and meeting space. Public phones ashore. Garbage disposal for customers.

Adjacent facilities: Covered barbecue pits in cabana on the waterfront. Firepit. Pet area. Play area. Cleaning facilities–sink and power. Club bookings, facility day use. Lawn Games.

Note:
Entrance depth to 7 1/2' at zero tides. Fuel dock accessible either side.
Covered docks at entrance are private.
Clear customs for USA destinations at Roche Harbor or Friday Harbor.

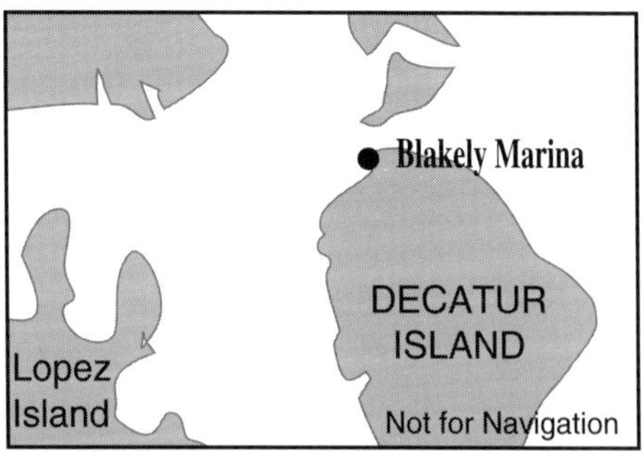

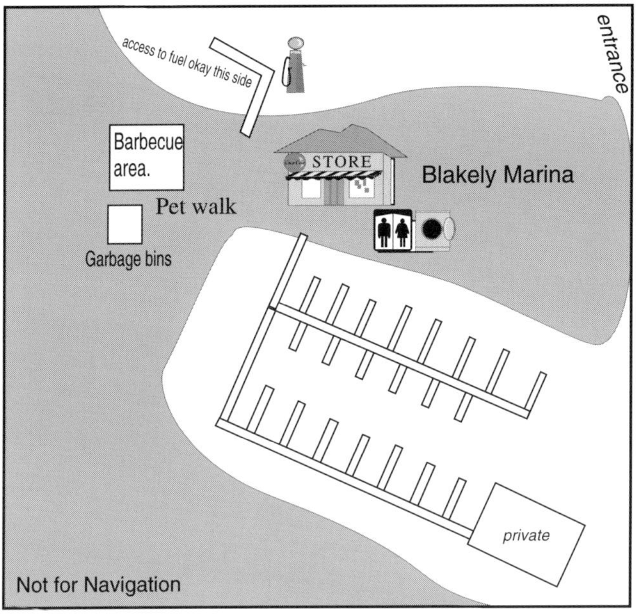

The entrance is narrow and shallow at low tide. The level of the water–drops to 8 feet but it is an easy passage.

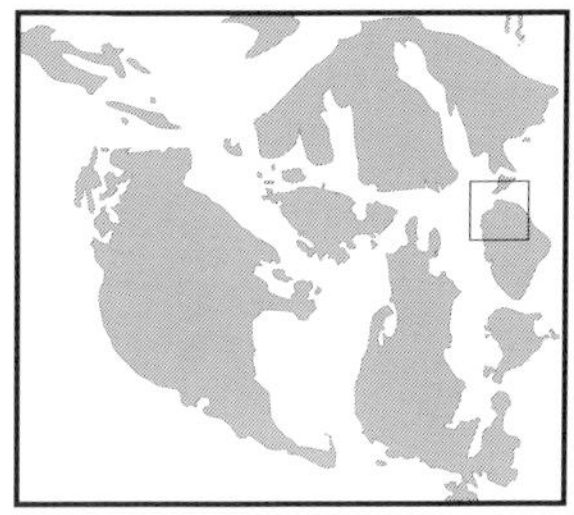

Left: The store at Blakely Marina with the patio out front. It faces the open passage and overlooks the fuel dock.
Below: The fuel dock is accessible from either side.

Blakely Marina

There is no access to the rest of the island, which is privately owned. The marina has facilities for guest boats and a sheltered bay where a number of boats can moor. Access is limited to higher tides for deep draft vessels. The depth through the channel is 8' at zero tides. A current runs past the entrance. This is a beautifully landscaped island and although it is private the scenery is enjoyable from the marina or the patio ashore.

Fisherman Bay

Islands Marine Centre

Ron & Jennifer Meng
Box 88, Lopez Island,
Washington. 98261
Phone: (360) 468-3377 Fax 468-2283
Chart 18421/30/34 VHF call channel 69
email: *imc@rockisland.com*

Marina services:

Moorage. Guest moorage on the south side of marina.
Power: 30 amp.
Water.
Fuel: Next door at Islander Lopez Marina.
Boatyard. Repairs and service. Haulouts.
15 ton travel lift. Launch ramp.

Customer services:

Showers, washrooms
Yacht sales and service. Various makes of outboard motors and boating equipment. Marine chandlery, hardware, charts, clothing, gifts, electronics, fishing tackle and information. Licences. NAPA store. Ice. Public phones ashore.
Walking: Road access walking, cycling.
VHF call 16, 69.

Entertainment: Golf nearby. Picnic and barbecue area. Bicycle rentals. Wine tasting at nearby vineyards.
Adjacent facilities: Lopez Village less than one mile. Stores–groceries, provisions, inn, meals, accommodations. Lopez Island Resort and restaurant next door.
Courtesy van service to ferry, airport, golf, by arrangement.

Note:
Customs Port of Entry at Friday Harbor or Roche Harbor.

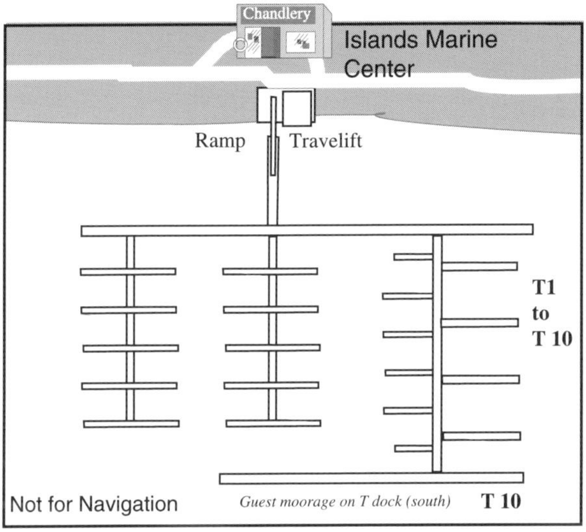

The entrance, above, to Fisherman Bay is narrow and shallow, but deep enough for a safe passage for most craft at low tides. A sign on marker number five reminds mariners to round marker number eight inside the bay leaving it to starboard. It is located not far from the outer floats of Islands Marine Center docks shown in the distance and to the right in the aerial photograph, opposite lower. The chandlery and service facility at Islands Marine Center appear in the photo below.

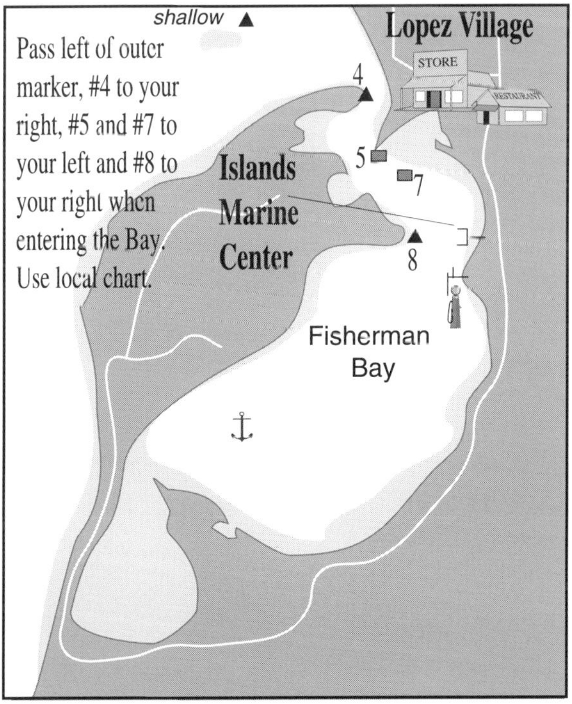

Islands Marine Centre

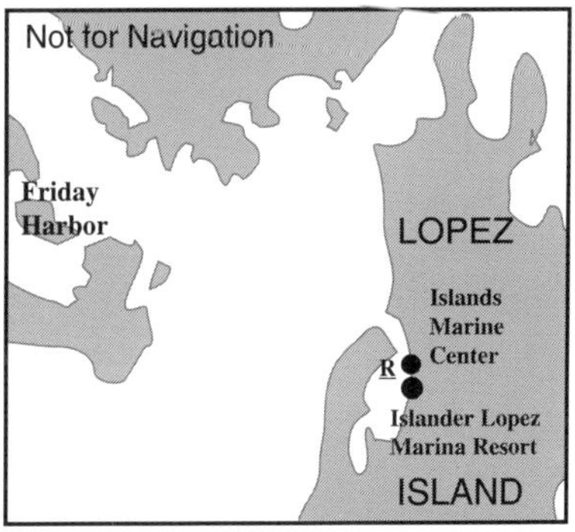

Lopez Islander Marina Resort

Box 459, Lopez Island,
Washington. 98261
Phone: (360) 468-2233 Fax 468-3382
Charts 18421/30/34
Dock 468-3383 VHF 78

Marina services:

Moorage. Guest mooring.
Power: 30, 50 amp.
Water.
Fuel: Gas, diesel
Sea plane float

Customer services:

Showers, laundry, washrooms.
Store on dock, groceries, provisions, charts, clothing, gifts, books, bread, fishing tackle. Ice. Public phones ashore.
Courtesy phone on dock.
Hotel accommodation–28 rooms. Swimming pool. Hot tub. Conference and banquet facilities. Salmon fishing and scuba diving charters. Wildlife & whale watching.
Restaurant (360) 468-2234. Patio service. Expresso bar on dock. Dockside welcome package June to September.
Walking: Road acccss walking, cycling.

Entertainment: Golf nearby.
Bicycle and kayak rentals and guides.
Adjacent facilities: Lopez Village less than one mile. Stores, arts and crafts, churches, accommodations. Near ferry service. Float plane calls at marina.

Lopez Island Vineyards are open to the public and wine tasting is available June to September 7, Wednesdays to Sundays from 12 noon to 5pm. Other months on Fridays and Saturdays. The vineyards are a short distance beyond the Lopez Island village and a reasonable walk or a quick drive away. Lopez wines are made from local Madeleine Angevine, Siegerrebe and from Yakima Valley Chardonnay and Cabernet Sauvignon/Merlot.
Lopez Island is one of the easiest islands in the San Juans for cycling. Bicycle rentals are available adjacent to Islander Lopez Marina and Islands Marine Center. An extensive road system allows wide exploration of the many interesting things to see and do on the island.

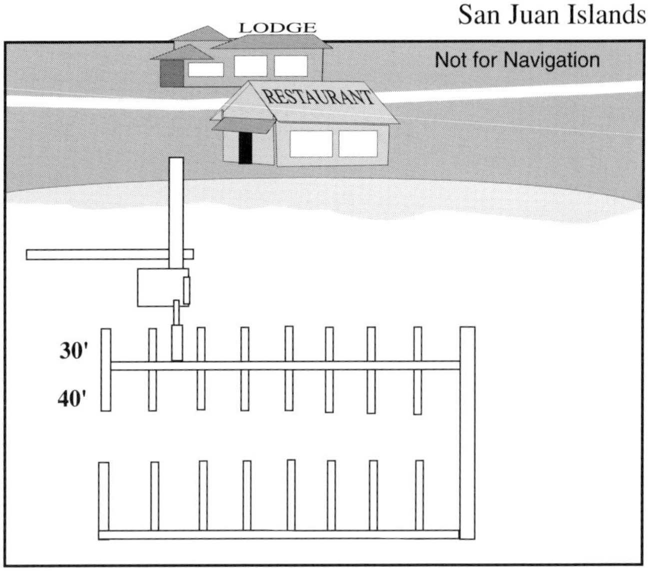

Clockwise from top: A view out of Fisherman Bay from the marinas. The fuel dock at Lopez Island Marina, two scenes at Lopez Village–a craft store and the village inn.

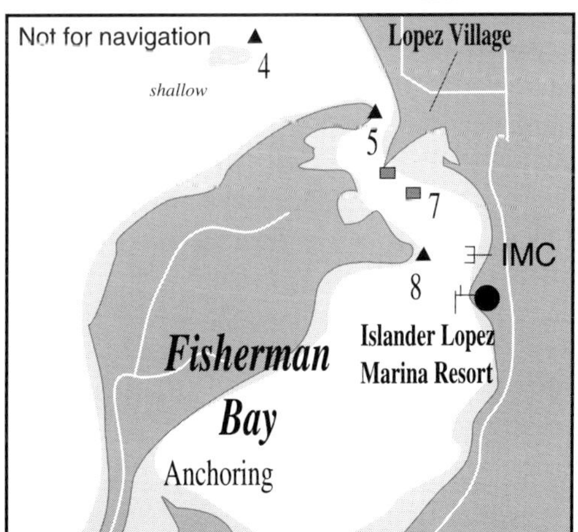

Point Roberts

Point Roberts Marina

Bruce Gustafson
713 Simundson,
Point Roberts, WA. 98281
Phone: (360) 945-2255 Fax (360) 945-0927
Charts: Cdn: 3450. US: 18400
email: *prmarina@whidbey.com*
Web: *www.prmarina.com\prmarina*

Customer Services:

Moorage 24' to 125'.
Fuel: gas, diesel, propane.
Water, Power 30, 50 amp, **Showers, laundry, washrooms.** Public phones. Restaurant. Chandlery, bait, tackle, fishing licences, repairs, service. Pump out facility at guest dock. Workyard. Haulouts–35 ton travelift, monorail sling hoist–boats to 22 feet.
US Port of Entry. Customs.
Mostly for permanent moorage. There is space for visiting vessels at the long finger in the entrance channel as well as other slips mostly on H dock. Phone for reservations.

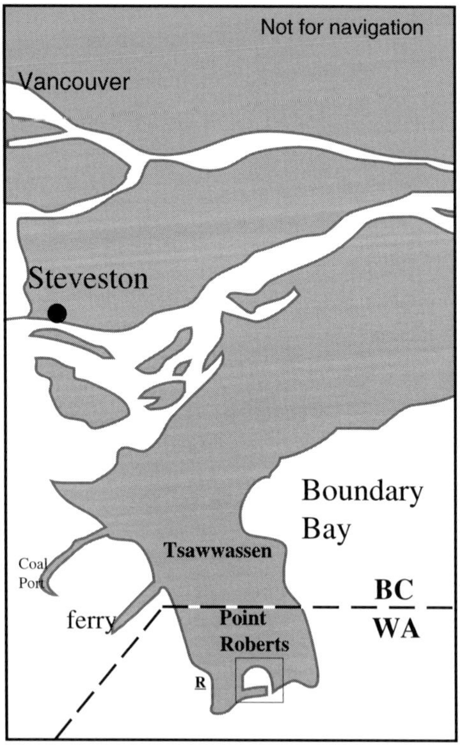

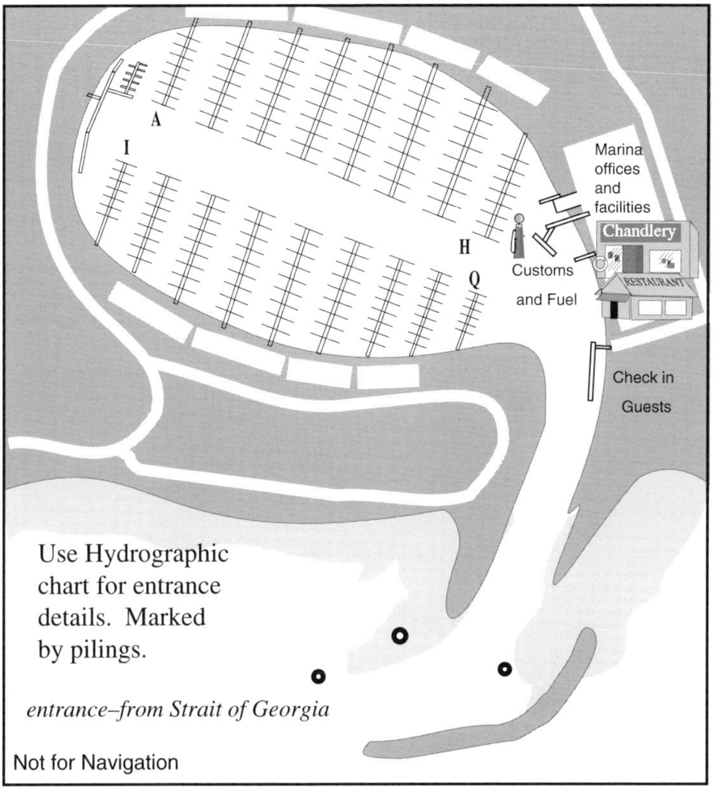

Although this marina is not in the San Juan Islands, it is very near. Access to it from the islands is closest from Sucia Island. It is a large centre for repairs and provides a good stop for access to the British Columbia mainland. Customs requires reporting when landing from Canadian ports. Point Roberts is on the tip of a peninsula which is attached to the BC mainland.

A nearby border crossing by land takes you into Greater Vancouver.

Coastal Marine Parks

(see ***Anchorages and Marine Parks*** *published 1998*)

Marine Parks in the San Juan Islands and British Columbia offer tranquil and delightful moorage. They have been established for use by the general public and attract hikers, backpackers, cyclists, RV campers and mariners. Many parks have picnic and overnight camp sites with trails and beaches. Some have docks adequate to moor dinghies only, some to moor a number of small to medium sized craft and others none at all. Most have mooring buoys for safe overnight mooring. Flat fees are levied for use of docks or mooring buoys. These change periodically but at present they are about $5 per mooring buoy per night and the charges usually apply from 3 pm to 8 am in the San Juans and after 6 pm in Canada.

The authorities ask that parks be respected and kept clean. Garbage should not be disposed of unless there is a specific disposal station. Sewage should not be discharged in marine park anchorages and noise should be limited to daylight hours. There are marine park hosts at some Marine Parks and in BC their presence will be indicated by the flying of a BC Parks Marine Park Host burgee. The host is usually a member of a power squadron, a yacht club, a sailing association or is an individual who has volunteered to assist visitors to the parks.

In the San Juans it is possible to purchase an annual moorage permit which allows use of mooring buoys and docks at the various locations throughout the year without further fee. Fees are in effect year round at some facilities while at others only between May and September. In the San Juan Islands these include Sucia Island, Stuart Island, Jones Island, James Island and Matia Island.

For more information on marine parks in US waters contact Washington State Parks Headquarters, PO Box 42664, Olympia, WA. 98504–2664 (360) 753-5771.
In Canada, RR #6, 2930 Trans Canada Hwy, Victoria BC. V9B 5T9. (250) 387-4363.

Vancouver Island
Southern Section

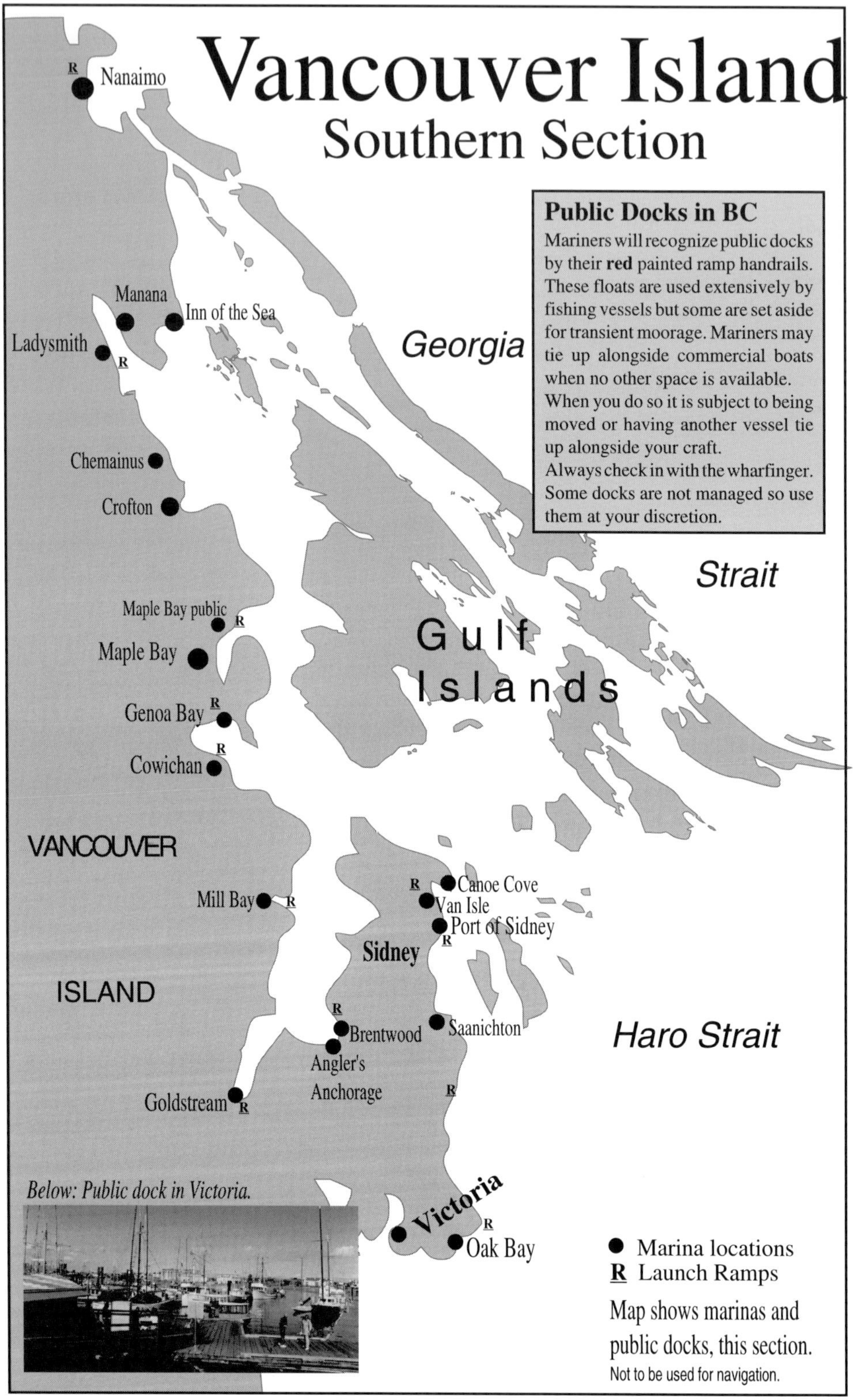

Public Docks in BC

Mariners will recognize public docks by their **red** painted ramp handrails. These floats are used extensively by fishing vessels but some are set aside for transient moorage. Mariners may tie up alongside commercial boats when no other space is available. When you do so it is subject to being moved or having another vessel tie up alongside your craft.
Always check in with the wharfinger. Some docks are not managed so use them at your discretion.

Below: Public dock in Victoria.

Map shows marinas and public docks, this section.
Not to be used for navigation.

Composite photograph above shows Victoria's Causeway floats in front of the Empress Hotel.

Victoria

There are many reasons to put into Victoria. It is a major customs port with a customs dock alongside the large public marina on Wharf Street. After clearing customs it is usually easy to find a slip at the public dock. Another public dock with limited space on either side of a single slip is located just before the Johnson Street bridge. Additional moorage under the same control as the Wharf Street dock is located right in front of the Empress Hotel. Beyond the Johnson Street bridge is The Gorge, and if you are ever in Victoria for a while a slow cruise up there. It is a worthwhile undertaking, just for the scenery.

West Bay has a number of floats and is the location of West Bay Marina with its narrow entrance and rock lined passage. A variety of services can be had at this marina but only occasional overnight moorage can be arranged for smaller vessels.

As far as Victoria is concerned there is not only a wide variety of stores but also the charm of an array of British styled shops, pubs and restaurants. British imports are the speciality and if the atmosphere of old England does not strike

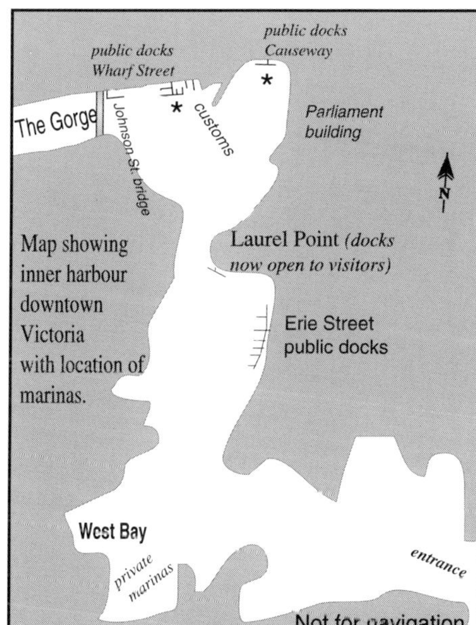

**Ph (250) 363-3273 up to 72 hours stay May 1–Sept 30.*

you immediately go and reserve afternoon tea at the Empress Hotel. Tourists flock to Victoria each summer and the attractions, apart from the above, include the the Royal BC Museum, parliament buildings, wax museum, undersea gardens, scenic London-bus tours and much more. There are festivities, shows and numerous public events in the city throughout summer.

Inset shows docks at Laurel Point. They are being expanded and opened to visitors.

Wharf Street floats

Victoria

Causeway Floats

Wharfinger
1002A Wharf Street,
Victoria, BC, V8V 1Y4
(250) 363-3760 Fax 363-6947
Chart 3313, 3415 VHF 73

Marina services:

Moorage: Transient moorage. Reserve in summer.

The Empress Hotel is located opposite the Causeway floats.

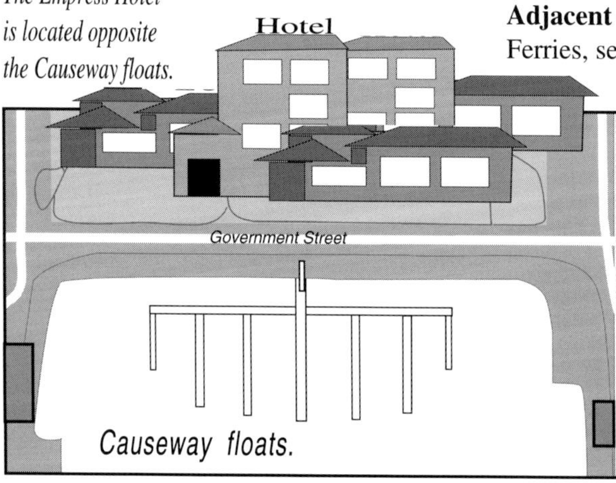

Water at dock.
Power: 30 amp.
Laundry, showers, washrooms.
Garbage disposal,

Customer services:
Customs/phone. Nearby city downtown within walking distance. Hotel Empress across the street. Walking–promenade.
Entertainment.
Downtown Victoria.
Adjacent or nearby services:
Ferries, seaplane services, customs. Hospital, banks, bank machines, car rentals, post office, swimming pool, telephone, taxi service. Throughout summer there are various forms of entertainment ongoing on the promenade and in the city. This is a busy, small marina and additional moorage is available at the Wharf Street dock.

Opposite: The docks at Wharf Street are probably the first place most mariners look for moorage. The alternative is in front of the Empress Hotel, which is very busy in summer, or at Erie Street, which is mostly occupied by commercial boats.

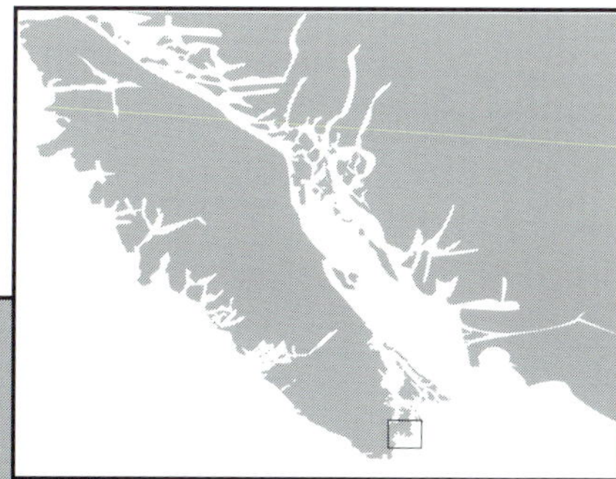

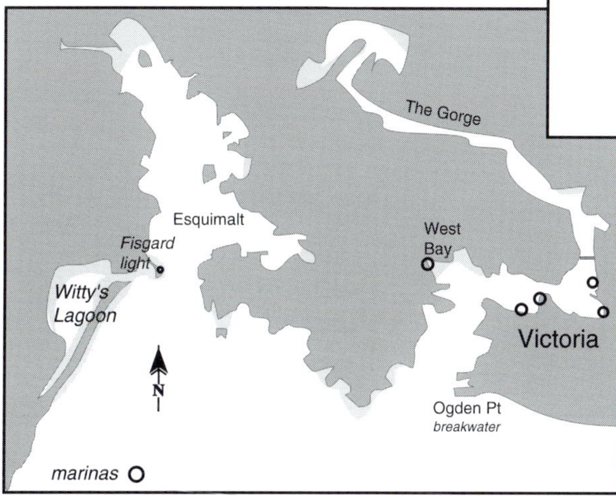

The diagram left shows Victoria Harbour in relation to the adjacent Esquimalt harbour where dock facilities are occupied by the Canadian military.

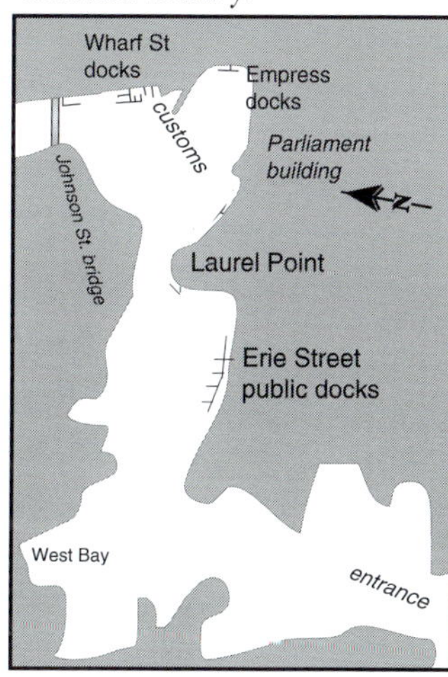

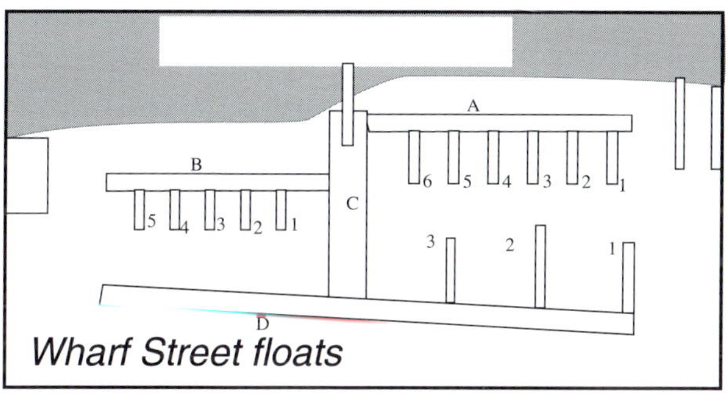

Wharf Street Floats

Wharfinger
1002A Wharf Street,
Victoria, BC, V8V 1Y4
(250) 363-3760 Fax 363-6947
Chart 3313, 3415 VHF 73

Marina services:

Moorage: Transient moorage.
Reserve in summer.
Water at dock.
Power: 30 amp.
Laundry, showers, washrooms.
Garbage disposal.

Customer services:
Customs/phone.
Nearby city downtown within walking distance.
Hotels nearby. Walking–promenade.
Entertainment.
Downtown Victoria, Old City blocks and Chinatown. Scuba diving at Breakwater.

Adjacent or nearby services:
Ferries, seaplane services, customs.
Hospital, banks, bank machines, car rentals, post office, swimming pool, telephone, taxi service.

Erie Street Public docks

Wharfinger
12 Erie Street, Victoria,
BC, V8V 1Y4.
Ph: (250)363-3723 Fax (250) 363-6947
Chart 3313, 3415 VHF 73

Marina services:

Fuel: (at adjacent fuel dock) gas, diesel. Mechanic and services available from local and nearby marine operators and facilities.
Moorage: Permanent moorage and overnight slips. Reserve in summer.
Water at dock.
Power: 20 amp
Garbage.
Laundry, showers, washrooms.
Waste oil disposal. Tidal Grid.

Nearby:

Accommodations, stores, banks, post office, dive shops, car rentals, boat rentals, groceries, hardware, restaurants, swimming pool.

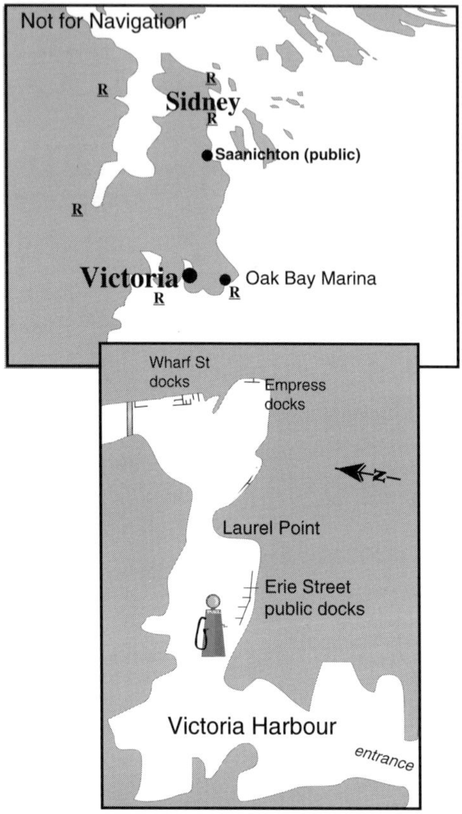

Saanichton, Sidney Public.
Transport Canada dock
Chart 3313, 3441, 3462
Manager • Float length 10 m
Lights •
Near Sidney restaurants, shops, medical, churches. (not walking distance).
Ferries nearby to BC mainland and US ports.

Oak Bay Marina

Todd Jakubowski
1327 Beach Drive, Victoria, BC,
V8S 2N4 (250) 598-3369
Phone: (250) 598-3366 Fax (250)598-1361
Chart 3313, 3462 VHF 68

Hazard: Rocky entrance. Shoals and shallows–marked with buoys. Consult charts.

Marina services:

Fuel: Available at adjacent fuel dock. Mechanic and services available from local and nearby marine operators and facilities.
Moorage: Permanent moorage and overnight slips. Reserve in summer.
Water at dock.
Power: 15 amp, 30 amp.
Laundry, showers, washrooms.
Security gate.

Customer services:
Customs/phone.
Nearby village. Oak Bay Avenue. 6–8 blocks. Chandlery. Tackle shop. Gift shops and others within walking distance.
Deli at marina. Restaurant–dining.
Walking road and beachfront.
Entertainment.
Nearby: Victoria. Shuttle service available.
Adjacent services:
Boat haulouts. Service. Launch ramp nearby.

Going north–a beginning

The fabled Gulf Islands are steeped in history and folklore and there are books on many subjects dealing with these fascinating islands as well as the San Juans. But this guide is not intended to keep you reading reams of historic notes, rather it encourages you to pick up the more complete works of other authors who have spent far more time researching those subjects for publication. As you make passage along the east shore of Vancouver Island, stopping possibly in Victoria or Sidney to clear customs, you unfold places of interest and people of charm and character that will please you. You may prefer to tie up at a dock and take in the local facilities. Try the Port of Sidney, Canoe Cove or Van Isle Marina. All three have good restaurants including places where you can sit and take in the magnificent crimson summer sunsets. Consider the Blue Peter at Tsehum Harbour, the Stonehouse Pub at Canoe Cove or any of a variety in Sidney including the hotel on the waterfront.

In the Victoria Harbour moorage options include the Causeway floats in front of the Empress Hotel. This location gives the visiting mariner the opportunity to see the best of the city. Victoria boasts one of the top tourism billings in Canada, with its very British atmosphere highlighted by pedigree pubs, London open-top buses and High Tea at the Empress.

In Saanich Inlet discover waterfront restaurants such as the Rusty Duck alongside the Brentwood ferry terminal, or the Brentwood Inn itself. Stop at Anchorage Marina and take your dinghy around to Butchart Gardens for a day of strolling in one of the most magnificent masterpieces of landscaping anywhere. Or pull into Mill Bay and walk up to the local shopping centre for some tastycappuccino or shopping at the well-stocked Thrifty Store. Continue up the coast through Genoa Bay for a safe overnight stop and fine dining, or visit Maple Bay for good moorage and a sumptuous meal at the marina restaurant.

Port Sidney

Port Sidney Marina

9835 Seaport Place,
Sidney, BC. V8L 4X3
or P.O. Box 2130,
Sidney, BC. V8L 2P6
Phone: (250) 655-3711
Fax: (250) 655-3771
Chart 3313, 3476, 3441, 3462
VHF 68
Web: www.

Marina services:

Fuel not available. Fuel docks at Tsehum Harbour and Canoe Cove (nearby, north) Mechanic and services available from local and nearby marine operators and facilities. Marine stores on dock and uptown Sidney.
Moorage: 300–400 slips. Permanent moorage and plenty of overnight slips. Reserve in summer. Excellent moorage. Also dockominium ownership available.
Water at dock. Taps at all fingers.
Power at docks: 50, 30, 20, 15 amp.
Laundry, showers, washrooms.
Activities dock–reserve for group private functions.

Customer services:

Nearby churches, post office, general stores, books, charts, fishing licences, tackle, bait, fresh produce, groceries, bakery, butchery, hardware, dry cleaning, liquor, pharmacy, clothing, gift stores all within walking distance.
Numerous restaurants, coffee shops, bistros and cafes.
Walking trails or road access.
Some beachfront walks.
Town streets and waterfront roads allow views while walking.
Kayak and other small craft rentals available. Scuba diving arrangements and charters, whale watching–ask marina for details.
Water taxi service to other marinas.

Entertainment.

Live music at marina in summer holiday months.
July 1 to 4: Fireworks display, dinner on the docks and live entertainment.
August: Model ship regatta.
Sponsored events all summer.
Nearby historic Butchart Gardens. Shuttle service in summer. Also to downtown Victoria, Airport, Ferries.
Sidney Museum summer hours–7 days per week. Features whales with murals and other historic exhibits.
Golf, tennis and other recreation nearby.

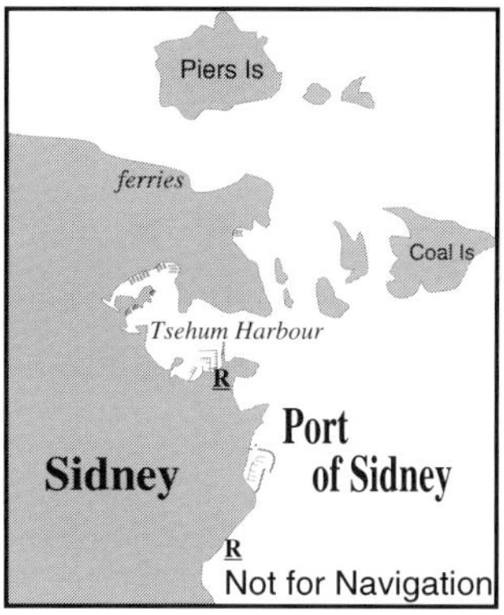

The Port Sidney Marina attracts many boaters from BC and Washington. It is a busy customs port and allows access to the many shops and services in the town.

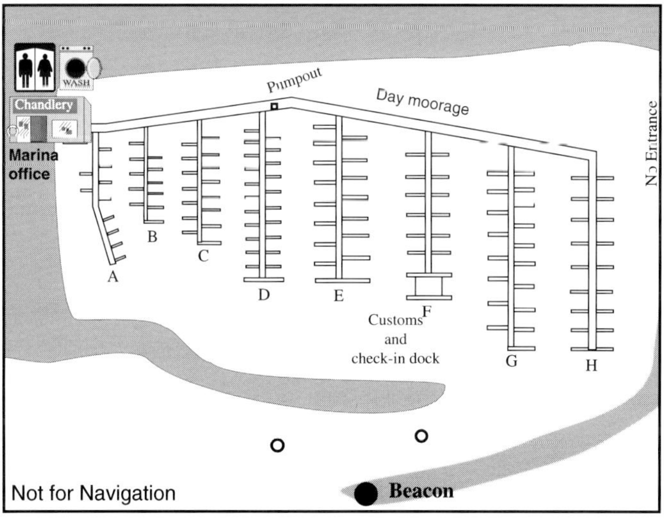

The aerial photograph above and the diagram left show the magnitude of the marina and the layout of the docks.

Tsehum Harbour

Chart 3313, 3476, 3441 VHF 68

Van Isle Marina

Mark Dickinson
2320 Harbour Rd., Sidney,
BC. V8L 3S6.
Phone: (250) 656-1138
Fax: (250) 656-0182
email: *info@vanislemarina.com*

Hazard–
Rocks in channel marked by beacons.

Customs services.

Designated marina for courtesy customs clearance. Customs dock. Phone on dock.

Marina entrance.

From around breakwater at entrance to Tsehum Harbour. Most visitor moorage is at east end of fuel and adjacent floats.

Marina services:

Guest Moorage

Fuel: Gas. Diesel. Stove oil: PetroCanada fuel dock. Outboard mix.

Water at dock. Multiple outlets.

Power: 50, 30, 15 amp. 120, 208 amp available. Public phone–ashore and on docks. Marine store on fuel dock. (250) 656-1138. Waste oil disposal, holding tanks pumpout, Ice. Charts, fishing tackle, licences, bait, life jackets etc.

Mechanic and services available from local and nearby marine operators and facilities. Haulouts. Philbrooks Shipyard within marina complex–vessels to 110 feet. Casual fine dining in Dock 503 Waterfront Cafe ashore. Yacht sales and service in marina complex. Boatlift within marina to 30 tons. New docks to accommodate craft to 185 feet. Large permanent marina with many overnight mooring slips. Many facilities ashore. Reserve in summer.

Customer services:

Laundry, showers, washrooms.

Nearby church/es, licenced restaurant/s. Nearby fine restaurants open 7 days a week. Walking trails or road access. Some beachfront walks. Town streets and waterfront roads allow views while walking. Scuba diving arrangements and charters–ask marina for details.

Looking northwards over Tsehum Harbour. The fuel dock at Van Isle Marina can be seen in the lower centre of the photograph. Westport Marina, upper right of bay, also offers some transient moorage.

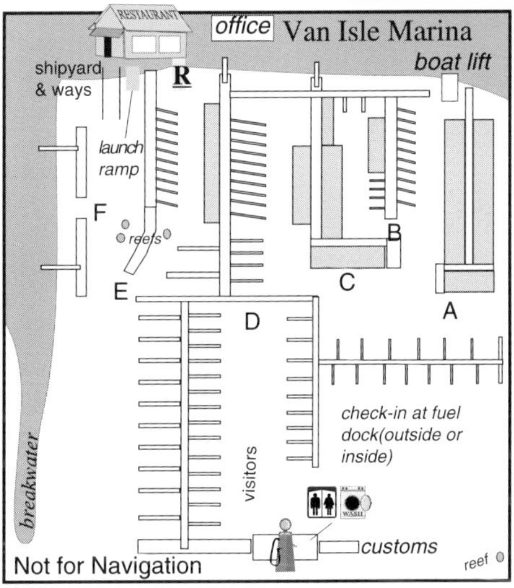

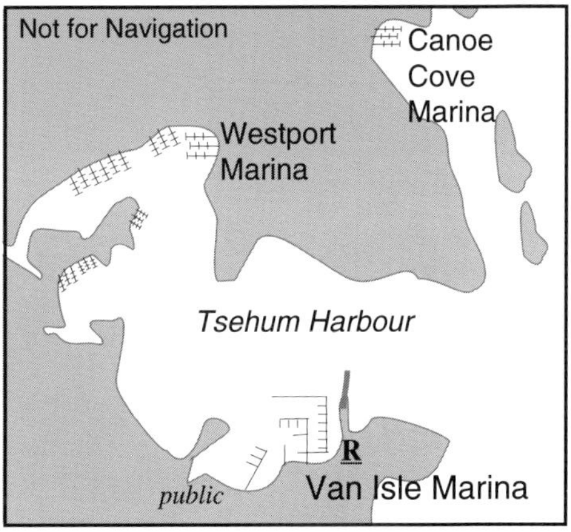

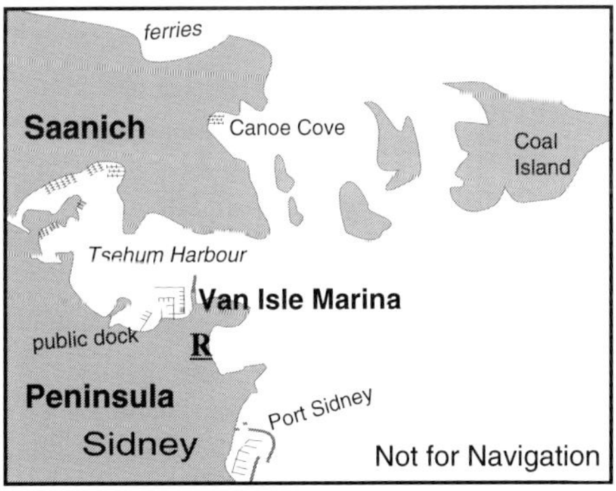

Tsehum Harbour
(Shoal Harbour Public)
Sidney, Vancouver Island.
Fish & Oceans
Chart 3310, 3476, 3441.
Managed • Float length: 318 m
Ramp • Breakwater • Garbage • Waste oil disposal • Parking • Water • Lights • Power • Telephone • Washrooms • Adjacent marinas, restaurants, chandlery, haul outs, marine repairs and full shipyard services.

Entertainment.

Cablevision available. TV and telephone hook-ups.
Nearby historic Butchart Gardens. Bus, taxi and rentals plus water shuttle to downtown Sidney, easy access to Victoria, airport, ferries. Golf, tennis and other recreation nearby.

Walk Roberts Bay Bird Sanctuary nearby.

Nearby facilities:

Philbrooks Shipyard, All Bay Marine, The Boat Yard, Jensen Marine, Sidney Marine Supply. Tanners Books.
Tsehum Harbour Government dock nearby.

Canoe Cove

Canoe Cove Marina

Chart 3313, 3476, 3441 VHF 68

Dock Manager
2300 Canoe Cove Rd. Sidney,
BC. V8L 3S6
Phone: (250) 656-5566
Fax: (250) 655-7197

Customs services.

Designated marina for courtesy customs clearance. Phone on dock.

Marina:
Some visitor moorage. Phone ahead to reserve overnight space.
Hazard: Rocks in north entrance marked by poles. Narrow waterway to fuel dock. Go Slow. Use right-of-way.

Marina services:

Fuel: Gas. Diesel. Stove oil. Propane. Ice.
Marine chandlery: Charts, fishing tackle, licences, bait, life jackets etc.
Mechanic and services available from local and nearby marine operators and facilities.
Haulouts. 35 ton travel lift. 65 ton ways.
Moorage: Large permanent marina with limited overnight moorage slips. Reserve.
Water at dock. Multiple outlets.
Power at docks: 50, 30, 20, 15 amp.
Laundry, showers, washrooms.

Customer services:

Customs dock/phone. Coffee shop. Pub and restaurant. Artist studio. Walking neighbourhood roadways. Scuba diving arrangements and charters–ask marina for details.
Public pay phones ashore.

Entertainment.

Nearby historic Butchart Gardens. Bus, taxi and rentals plus water shuttle to downtown Sidney, road access to Victoria, airport.
Golf, tennis and other recreation nearby.
Adjacent facilities: (The Stonehouse Pub–phone 656-3498) open 7 days a week.
Nearby: BC Ferries Swartz Bay to Vancouver and Gulf Islands.
Churches in sidney and nearby.

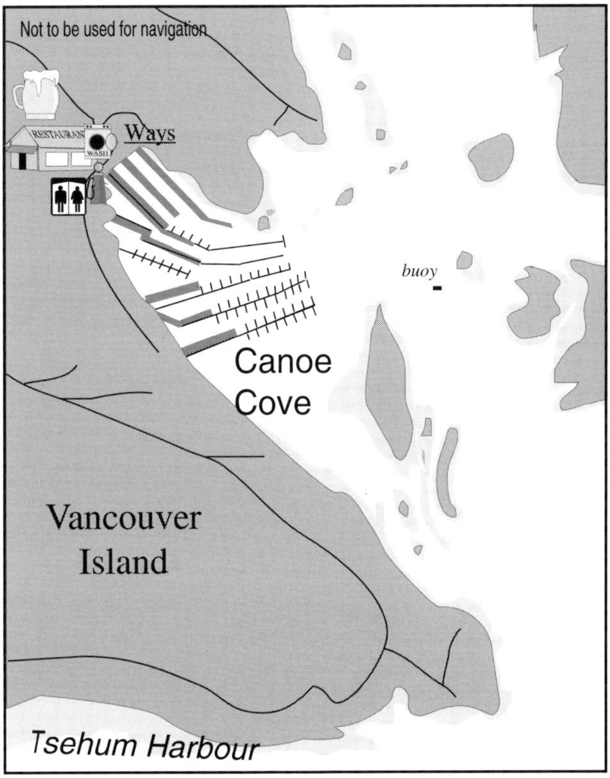

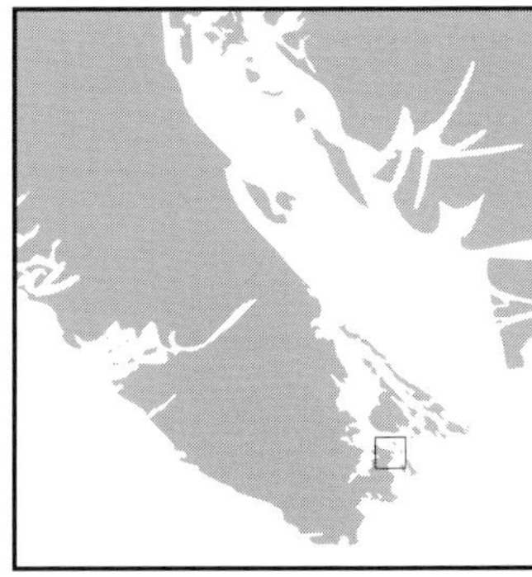

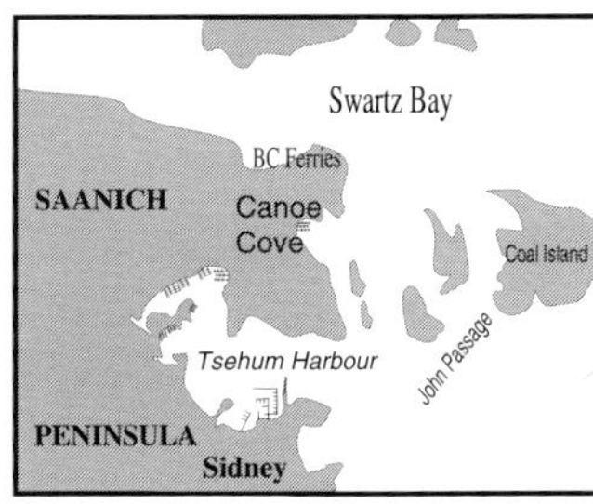

Canoe Cove

This cosy marina is a classic, with many permanent resident boats and little room for guests. Mariners stop in for fuel at the dock tucked away between docks C and D or to clear customs. There is a homey cafe for breakfast or lunch and early light dinner in the summer. For finer fare the Stonehouse Pub a short way behind the marina has an atmosphere that matches its good dining. Special events are celebrated at the restaurant with appropriate meals for the occasion.

The large boat yard adjacent to the marina has haulout and dry storage space and is operated by the service centre at the head of dock C. The ways will allow haulouts of larger vessels and the travel lift caters to more average sized boats. A chandlery situated near the cafe carries a wide range of items for repairs, service and annual maintenance.

The property is quite extensive with a resident artist and a well-known yacht brokerage company. It is the home of the Canoe Cove shipyard.

Many boat owners at the marina commute by ferry from the mainland and other parts taking advantage of the marina's close proximity to some of the most favoured cruising destinations on the coast and in the nearby islands.

Piers Island, Sidney (public)
Transport Canada dock
Chart 3313, 3476, 3462, 3441
Manager • Float length 63 m
Used primarily for local residents' access.
Opposite BC Ferries Swartz Bay terminal (Sidney).

Brentwood Bay

Brentwood Inn Resort

Mike Keepence
7172 Brentwood Drive, Brentwood Bay, BC. V0S 1A0
Phone: (250) 652-2413
Fax: (250) 652-2402

Customs services.

At adjacent Anglers Anchorage Marina. Phone on dock.

Marina Entrance.

Hazard–Reef off outer dock. Best route keep marker to starboard. Approach dock from ferry terminal. Check your chart.

Marina services:

Convenience store: Ice, fishing tackle, licences, bait, some provisions.
Mechanic and services available from local and nearby marine operators and facilities can be arranged.
Moorage: 65 slip permanent marina with limited overnight moorage. Reserve.
Water at dock. Multiple outlets.
Power at docks: 30, 15 amp.
Laundry, showers, washrooms.

Customer services:

Nearby church/es, Dockside restaurant and pub with view. Deck open in good weather and restaurant open 7 days a week.
Restaurant, speciality–seafood.
Walking trails or road access. Some beachfront and parkland trails.
Professional fishing guides available.
Rental runabouts - 16 footers.
Scuba diving arrangements and charters–ask marina for details.
Public pay phones ashore.

Entertainment.

Nearby historic Butchart Gardens.
(Dinghy in to Butchart Gardens)

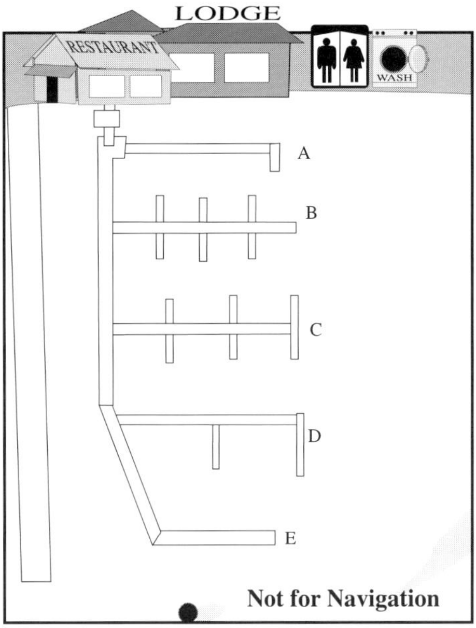

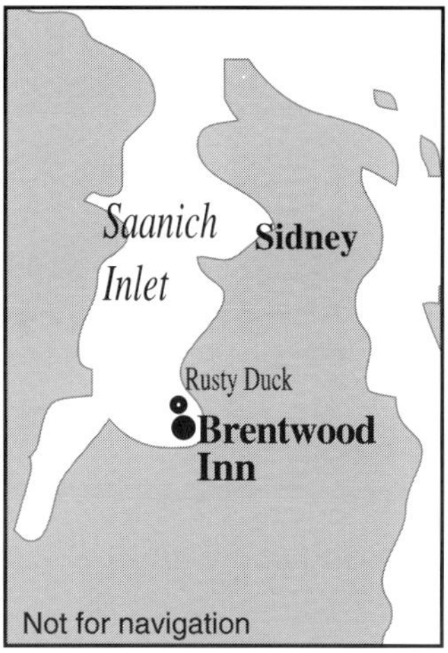

Charts 3313, 3441

Opposite: Brentwood Inn marina docks.
Above: The dock at the Rusty Duck is tucked in north of the ferry landing adjacent to Brentwood Inn. Small to medium boats only and it is shallow at low tide.

Bus, taxi and rentals to downtown Sidney, easy access to Victoria, airport, ferries. Golf, tennis and other recreation nearby.

Adjacent facilities:

Rusty Duck Restaurant–dock for restaurant customers.
Fuel at Mill Bay, Goldstream or Sidney.

Public dock

Brentwood Bay

Saanich Inlet
Transport Canada dock
Chart 3313
Manager • Float length 22 m
Adjacent: Private marina and Angler's Anchorage Marina
Walk in adjacent park.
Used mostly by commercial traffic for loading.

Angler's Anchorage

Angler's Anchorage Marina

Mark Tigchelaar
933 Marchant Rd. RR #1,
Brentwood Bay, BC. V8M 1B5
Phone: (250) 652-3531
Fax (250) 652-9923

Canada Customs

Designated marina for courtesy customs clearance.
Phone on dock.
Marina Entrance.
Near Brentwood Inn Marina–refer to chart 3313 for navigation.

Marina services:

Mechanic and services available from local and nearby marine operators.
Repairs can be arranged.
Moorage: Large permanent marina with limited overnight moorage slips.
Water. Power at docks: 30, 15 amp.

Chart 3313, 3441
VHF 68

Customer services:

Laundry, showers, washrooms.
Nearby church/es.
Dockside marine restaurant.
Public pay phones ashore.

Entertainment.

Nearby historic Butchart Gardens.
Bus, taxi and rentals to downtown Sidney, easy access to Victoria, airport, ferries.
Walking trails or road access. Some beachfront walks.
Golf, tennis and other recreation nearby.

Adjacent facilities:

Brentwood Inn Resort/Marina.
Government dock. Restaurants.
Shopping in nearby Brentwood Bay.

Pictured above: The dock and fuel pump at Goldstream. View looks north towards entrance of Saanich Inlet.
Opposite: Angler's Anchorage Marina.

Goldstream

Goldstream Boathouse

Mark Aitken and Doug Spence
2892 Trans Canada Highway, RR 6,
Victoria, BC. V9B 5T9
Phone: (250) 478-4407
Chart 3313, 3441 VHF 68

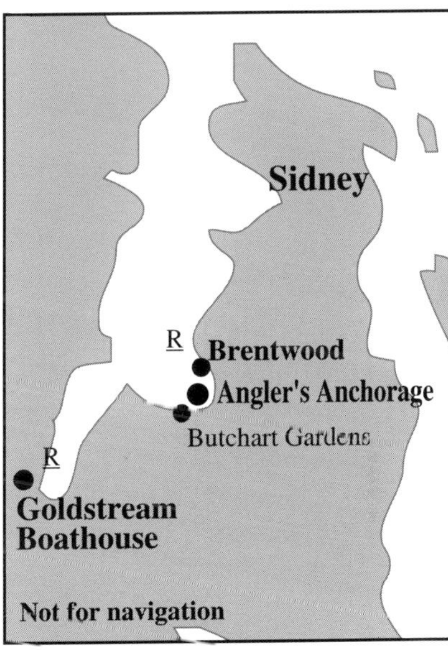

Marina services:

Fuel: Gas. Oil filters, etc.
Marine supplies. Marine repairs. Haulouts.
Moorage. Transient available.
Water at dock.
Power at docks: 30, 15 amp. (For transient moorage customers–15 amp.) **Washrooms.**

Customer services:

Nearby church/es
Road access walking. Parkland trails at Goldstream Park nearby.
Rental boats. Bait, ice, tackle, fishing licences. Groceries, snacks.
Scuba diving arrangements and fishing charters–ask marina for details.
Public pay phones ashore.

Entertainment.

Historic landmark boathouse.
Butchart Gardens.
Bus, taxi and car rentals to downtown Victoria. Golf nearby.

Adjacent facilities:

Campsites. **Launch ramp.**

Mill Bay

Mill Bay Marina

Fred and Marilyn Laba
Box 231 740 Handy Road,
Mill Bay, BC. V0R 2P0
Phone: (250) 743-4112

Marina services:
Fuel: Gas. Diesel. Launch ramp.
Water at dock. Multiple outlets.
Power at docks: 15 amp. Multiple outlets.
Laundry, showers, washrooms.
Boats for rent. Charters. Bed & breakfast.
Marine supplies. Ice, fishing tackle, licences, bait.
Mechanic and services can be arranged.
Moorage: 158 plus slips–permanent marina with 540' overnight moorage. Reserve.

Customer services:
Nearby church/es. Road access walking. Some nearby river and beach trails.

Chart 3313, 3441 CB 10

Rental boats. **Launch ramp**
Shopping centre and grocery store nearby (Thrifty) will deliver purchases to your boat.
Public pay phones ashore.
Entertainment.
Across inlet to Butchart Gardens.
Bus, taxi and rentals to downtown Victoria.
Golf nearby.

Mill Bay, Vancouver Island
Fisheries & Oceans public dock
Phone (250) 743-9764
Chart 3313, 3441, 3462
Manager • Float length 15 m

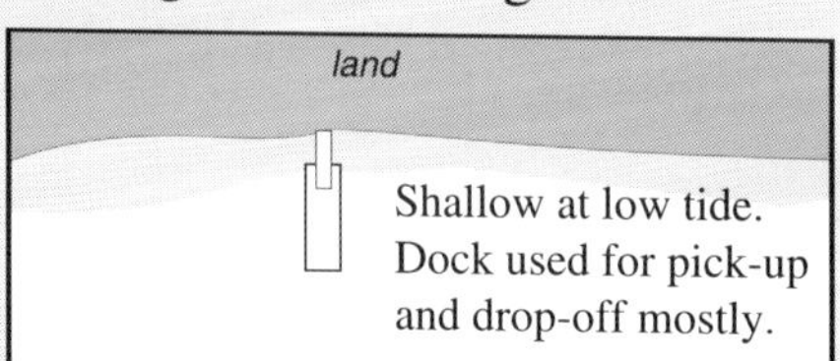

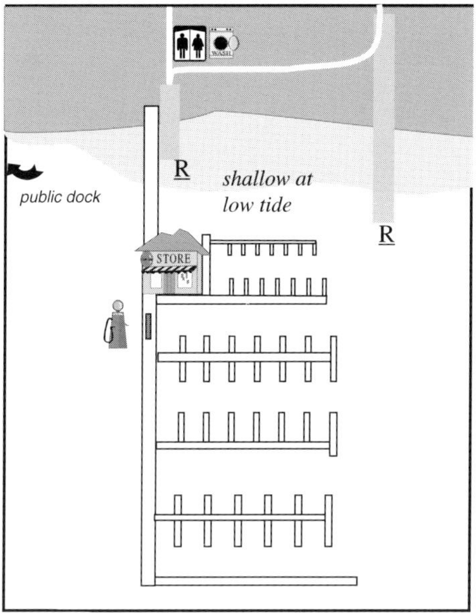

Not for Navigation

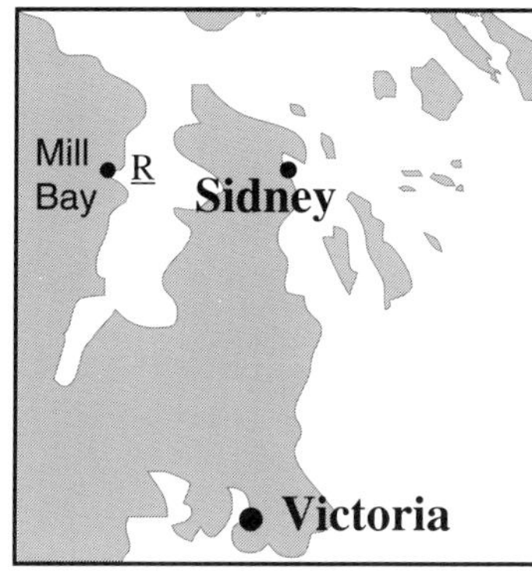

Mill Bay Marina

The shopping centre at Mill Bay has good grocery shopping as well as a variety of interesting stores, arts and crafts, restaurants, espresso and coffee shops among others.
The centre is an easy walk from the marina. The main float to the south includes a fuel dock and overnight moorage for larger vessels. Small craft will be assigned slips inside.

Right: Public docks. Left: Masthead Marina.

Cowichan

Numbers identified in photo on page 64

1. Cowichan Bay Public

Harbour Authority/Wharfinger
General Delivery,
Cowichan Bay, B.C. V0R 1N0
Phone: (250) 746-5911 Fax 748-7122

Marina services:
Water at dock. Multiple outlets.
Power at docks: 15 amp. Multiple outlets.
Washrooms.
Telephone.
Moorage: Overnight available. Check in on arrival or reserve.
Nearby services:
Showers, laundry, boats for rent. Charters. Bed & breakfast. Marine supplies. Ice, fishing tackle, licences, bait. Boat lift, liquor store, post office, bank machine, pub, restaurant, shops, swimming pool, taxi, hardware, grocery.

Chart 3313, 3441 CB 10

Mechanic and services can be arranged. Nearby church/es. Road access walking. Some nearby river and beach trails.

2. Masthead Marina

1705 Cowichan Bay Road,
Cowichan Bay, B.C. V0R 1N0
Phone: (250) 748-5368 Fax 748-5855

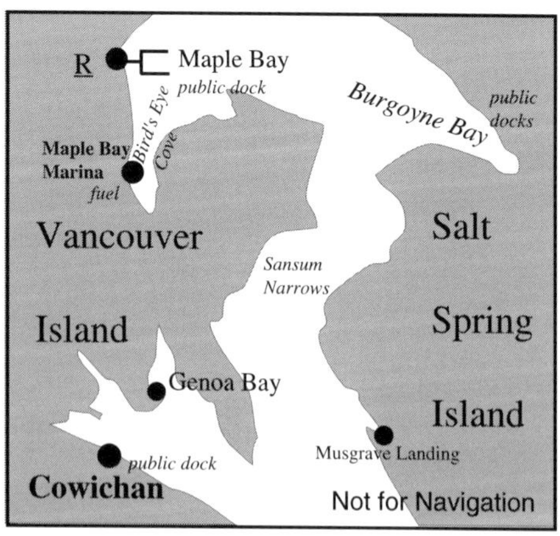

Masthead Marina left of jetty.

Looking west from the public docks

Marina services:

Garbage disposal.

Water at dock. Multiple outlets.

Power at docks. Multiple outlets.

Restaurant.

Moorage: Overnight available. Check in on arrival or reserve.

Nearby services:

Showers, laundry, washrooms, boats for rent, charters, accommodations. Marine supplies. Ice, fishing tackle, licences, bait. Boat lift, liquor store, post office, bank machine, pub, restaurant, shops, swimming pool, taxi, hardware, grocery. Art gallery, potter, jeweller.

Mechanic and services can be arranged.

Nearby road access–walking. Some nearby river and beach trails. Kayak and mountain bike rentals.

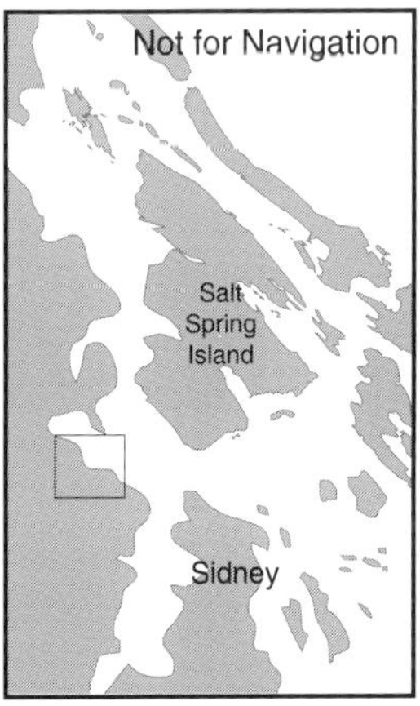

Update Information courtesy Sue Wells.

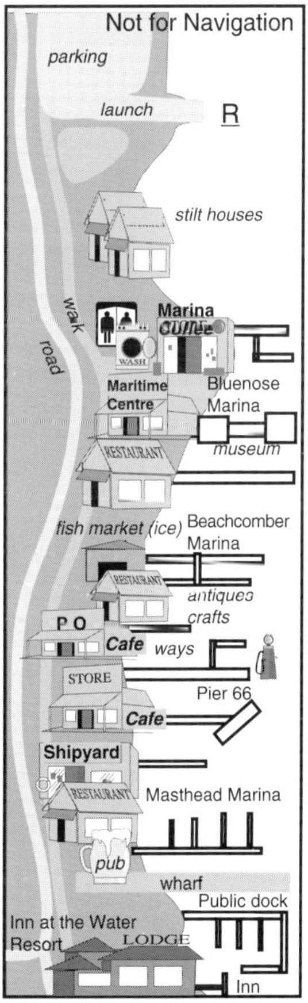

Cowichan Harbour is a working wharf. Visitors are welcome to stop and experience its ambience. The atmosphere of the village is captivating.

No need for typical gift stores and tourist boutiques. Moor at the public docks when the fishing fleet is away. And at other marinas in the Bay subject to space being available.

Photo opposite page shows Masthead Marina (left) and the public dock (right).

3. Pier 66 Marina

1745 Cowichan Bay Road,
Cowichan Bay, B.C. V0R 1N0
Phone: (250) 748-8444 Fax 748-8444
email: *rufus@islandnet.com*

Marina services:
Fuel: gas, diesel, premix, ice, oil, bait, licences. **Garbage disposal.**
Water at dock. Multiple outlets.
Power at docks: 15 amp. Multiple outlets. Telephone. Marine supplies, boat rentals, charts, fishing licences, bait and tackle.
Moorage: Overnight available. Check in on arrival or reserve.

Nearby services:
Showers, laundry, charters. accommodation, boat lift, post office, bank machine, pub, restaurant, shops, swimming pool, taxi. Tidal grid. Mechanic and services can be arranged. z Nearby church/es. Road access walking. Some nearby river and beach trails.

4. Beachcomber Marina

1759 Cowichan Bay Road,
Cowichan Bay, B.C. V0R 1N0
Phone: (250) 748-6789

Marina services:
Water at dock. Multiple outlets.
Power at docks: 30 amp. Multiple outlets.
Garbage disposal.
Washrooms.
Boat lift, telephone.
Moorage: Overnight available. Check in on arrival or reserve.

Nearby services:
Showers, laundry, charters. Bed & breakfast. Ice. Marine supplies, hardware, groceries, charts, tackle, boat rentals, fishing licences, bait. liquor store, post office, bank

machine, pub, restaurant, shops, swimming pool, taxi.
Mechanic and services can be arranged. Nearby church/es. Road access walking. Some nearby river and beach trails.

5. Bluenose Marina

1765 Cowichan Bay Road,
Cowichan Bay, B.C. V0R 1N0
Phone: (250) 748-2222 Fax 748-8982
Marina services:
Water at dock. Multiple outlets.
Power at docks: 15 amp. Multiple outlets.
Washrooms. Laundry. Shower.
Garbage disposal.

Telephone, restaurant, taxi.
Moorage: Overnight available. Check in on arrival (plus 5 mooring buoys).
Nearby services:
Boats for rent. Charters. Accommodation. Marine supplies. Ice, fishing tackle, bait, fishing licences. Boat lift, liquor store, post office, pub, shops, swimming pool, taxi, hardware, grocery.
Mechanic and services can be arranged.
Road access walking.
Some nearby river and beach trails.

6. Inn at the Water Resort

Howard Johnson Coast Inn Resort
1681 Botwood Road,
Cowichan Bay, B.C. V0R 1N0
Phone: (250) 748-6222 Fax 748-7122
Small dock located adjacent to entrance of the public marina. Hotel overlooks docks.
Marina services:
Water at dock.
Power at docks: Multiple outlets.
Washrooms, showers.
Telephone. Ice.
Car rental. Accommodation. Swimming Pool. Restaurant. Pub. Taxi service.
Moorage: Overnight available. Check in on arrival or reserve.
Nearby services:
Laundry, boats for rent, launching, charters. Bed & breakfast.
Marine supplies. Fishing tackle, licences, bait. Boat lift, liquor store, post office, bank machine, shops, hardware, grocery.
Mechanic and services can be arranged.
Nearby church/es. Road access walking.
Some nearby river and beach trails.

7. Cherry Point

1241 Cherry Point Drive,
Cobble Hill, B.C. V0R 1L0
Phone: (250) 748-0453
Marina services:
Garbage disposal.
Water at dock.
Power at docks: 30 amps.
Washrooms.
Telephone. Camping, boat launch, boat charter. RV facilities.
Moorage: Overnight available. Check in on arrival or reserve.
Nearby services:
Mechanic and services can be arranged.
Road access walking.
Cherry Point Marina is located on the approaches to Cowichan.

Genoa Bay

Genoa Bay Marina

Will Kiedaisch
RR #1 5100, Genoa Bay Road,
Duncan, BC., V9L 1M3
Phone: (250) 746-7621
Fax: (250) 746-7621

Hazard: Entering bay keep to port of day beacon. Avoid shore-side reef–marked.

Marina services:

Moorage. Large marina with permanent and transient moorage.
Water at dock. (Please conserve)
Power at docks: 15 amp.
Laundry, showers, washrooms.
Ice. Repair service available.

Customer services:

Restaurant. Licenced.
Breakfast, lunch, dinner. Also patio service. Barbecue area. Tenting.

Charts 3313, 3441, 3442, 3478 VHF 68

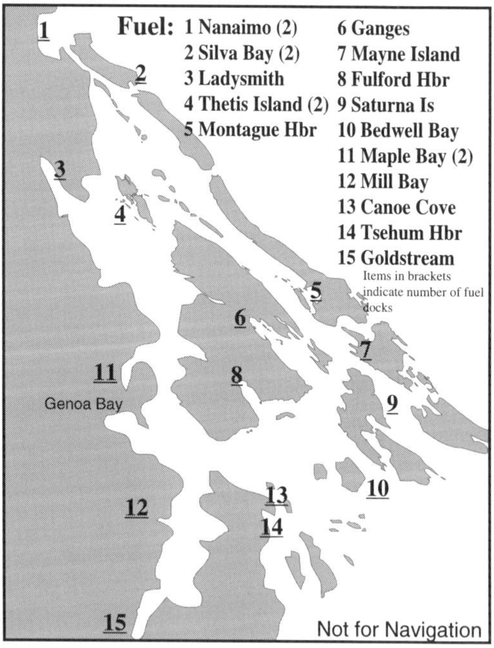

Above: Genoa Bay Marina in the spring.
Left: The restaurant sits on the bank overlooking the marina.
Opposite is an aerial view of the bay.

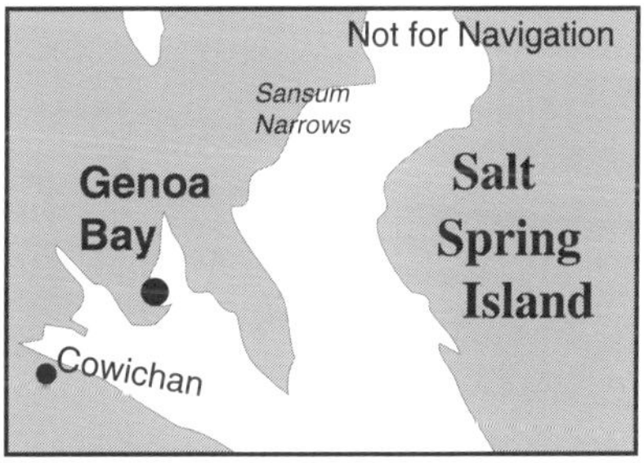

General store: groceries, fishing gear, licences, fresh baked goods. Gifts.

Art and crafts gallery.

Entertainment:

Road access walking or cycling. Nearby hiking trails.

Public pay phones ashore.

Adjacent facilities:

Anchorage. Yacht Sales. **Launch ramp**.

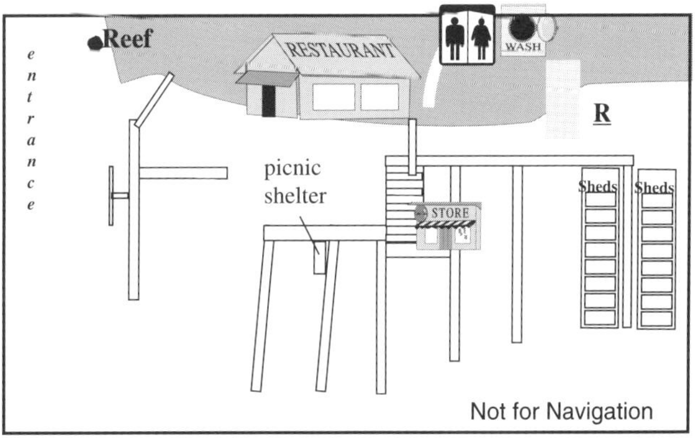

Maple Bay

Maple Bay Marina

Ted McLeod & Bill Neapole
6145 Genoa Bay Road,
Duncan, BC., V9L 1M3
Phone: (250) 746-8482
Fax: (250) 746-8490

Charts 3313, 3478, 3442 VHF 68
3441, 3462

Marina services:

Fuel: Gas, diesel. Marine repairs, service. CNG. Propane. Maple Bay Marine Centre (Jim Young & Ron Goss).
Marine supplies. Diving, towing.
Fishing gear, licences, charts, bait. Ice.
Guest and permanent moorage.
Water at dock.
Power at docks: 15, 20 amp.
Laundry, showers, washrooms.

Customer services:

Shipyard Pub & Restaurant. Licenced.
Patio service.
Gift Shop. General Store. Beauty salon.
Road access walking or cycling trails.
Scuba diving arrangements and charters–ask marina for details. Kayaking.
Public pay phones.
Daily float plane service.

Entertainment.

Restaurant. Outdoor breakfast adjacent to marina buildings available on weekends.

Adjacent facilities:

Excellent bed & breakfast accommodations nearby. Custom yacht work available.

Maple Bay

Fisheries & Oceans dock
Phone (250) 746-7101
Chart 3310, 3441, 3462
Manager • Float length 46m
Water • Lights • Public phone •
Launch ramp

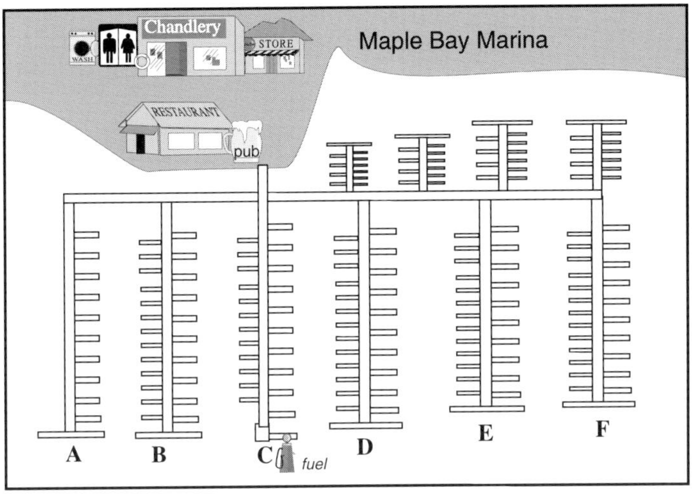

The public float, (left) at Maple Bay is in the outer bay before approaching Bird's Eye Cove. The general store at the government float was not operating in 1998. Lower right: Models displayed at the annual Classic and Wooden Boat festival at Maple Bay.

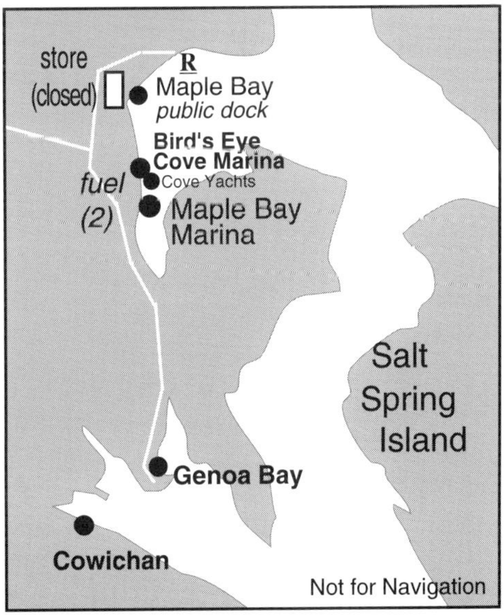

Birds Eye Cove Marina (left) is a popular fuel stop. Service and repairs at adjacent Cove Yachts. Maple Bay Marina, (previous page) at the head of the bay, is a major moorage and fuel stop with available marine supplies.

Bird's Eye Cove Marina

Bob Walker

6271 Genoa Bay Rd, RR1 Duncan, BC. V9L 1M3.

Phone (250) 748-4255 Fax (250) 748-2319

Marina Services:

Some transient moorage.

Fuel dock–all fuels and oils. Tackle, some supplies.

Service and repairs adjacent.

Open year round.

Nearby:

Maple Bay marina–store, restaurant, pub.

Cove Yachts (1979) Ltd.

Phil and Alexandra Pidcock

Cove Marina, 6261 Genoa Bay Road

Duncan BC V9L 5Y4

(250) 748-8136 Fax 748-7916

Travel lift. Ways. Commercial and pleasure craft work. Marine supply store.

On the floats at Maple Bay Marina during the Wooden Boat Festival. Above, right: Models on display at the festival.

Crofton, Chemainus, Ladysmith

An overnight stop at the sheltered harbours of Crofton and Chemainus can be a memorable experience. The government docks have room for transient boats in the summer. Off season if there is no room to tie up it is possible, preferably for not too long a stay, to come alongside a docked fishing boat. Fishermen generally do not object to having a boat moored temporarily alongside them. However you may prefer to be tied directly to the dock for easier access to and from your moored boat.

Top: Crofton docks are emptier of fish boats in summer than winter, providing lots of guest moorage. Above: Chemainus looks busy but there is often space to tie up. A new marina is being built nearby.

At Ladysmith, within easy walking access from the government docks, one can find many store facilities and boating requirements, supplies and services and a touch of coastal history. It is an old coal mining town named for the town in South Africa which was under siege during the Boer War and relieved by the British coincidental to the founding of Ladysmith in BC. Its naming also honoured the charitable wife of Sir Harry Smith, governor of the Cape Province of South Africa.

Across the harbour the docks at Manana Lodge are most hospitable to recreational boats. During summer a steady stream of craft call at the lodge for overnight moorage, fuel and very limited supplies. Also a regular clientele check in for the sumptuous meals served at the lodge. Moorage is sheltered and the docks are available specially for visiting boats.

From Ladysmith it is easy and quick to access several outstanding destinations. Among them are Telegraph Harbour, Wallace Island, Chemainus and Pirate's Cove. Some interest-

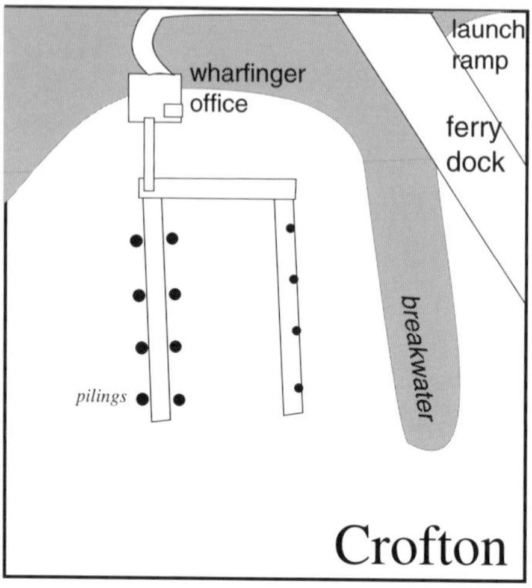

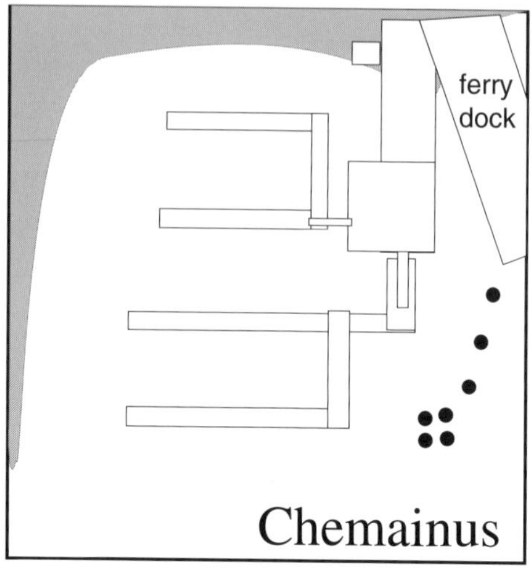

Public docks Crofton to Ladysmith

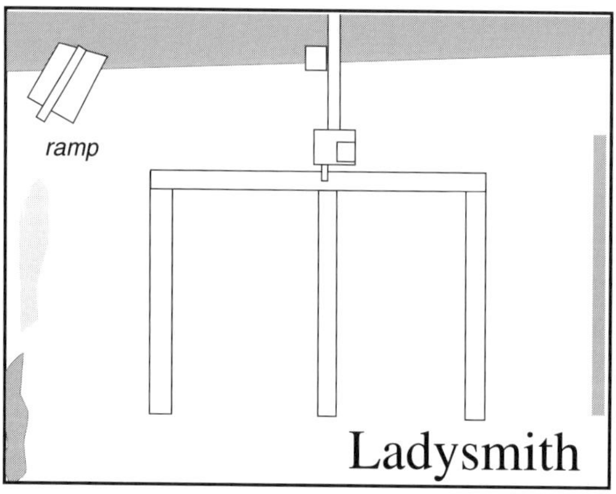

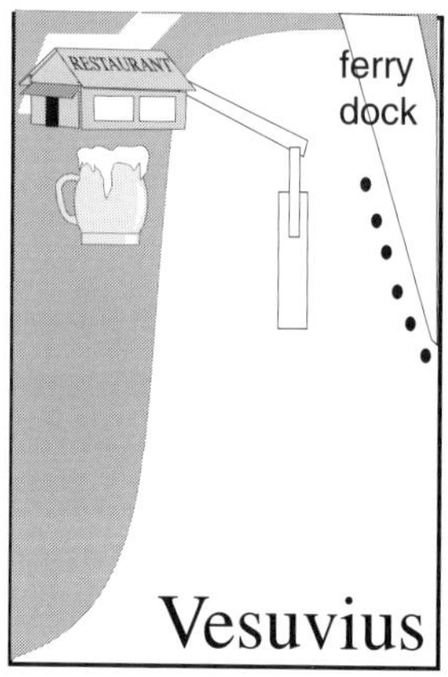

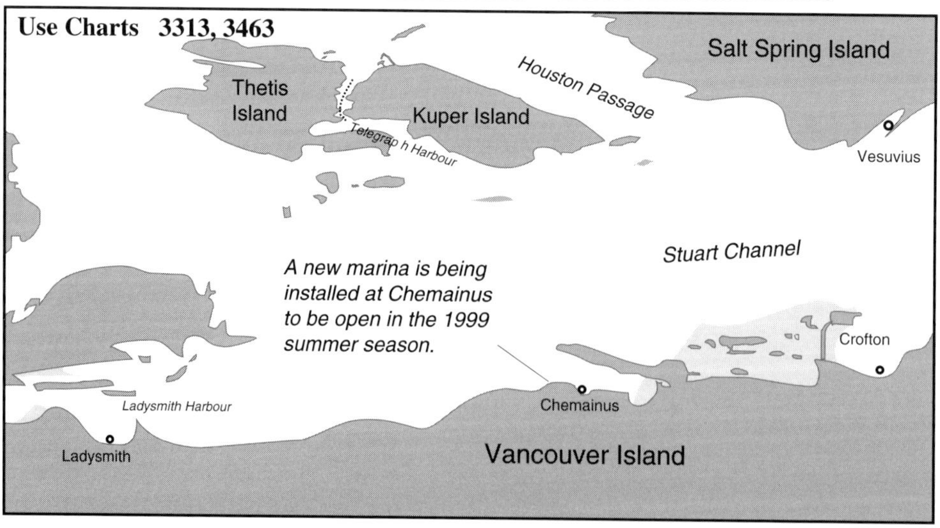

ing stops in this general area include North Galiano, Vesuvius, Crofton and Retreat Cove.

Visit Wallace Island Marine Park nearby for its excellent paths and scenery, for its history and protected overnight moorage. It's a cross-roads for people on the move, enjoying the outdoors by boating, canoeing and tenting. The park attracts many boats every season and many campers take advantage of the recreational facilities provided by the Parks Branch and maintained by Power and Sail Squadron members.

Above: Ladysmith public marina. Note location of launch ramp (lower left).
Above: Ladysmith Harbour from the air.
Below: In summer the docks at Ladysmith are often quite empty with lots of room to manoeuvre between fingers.

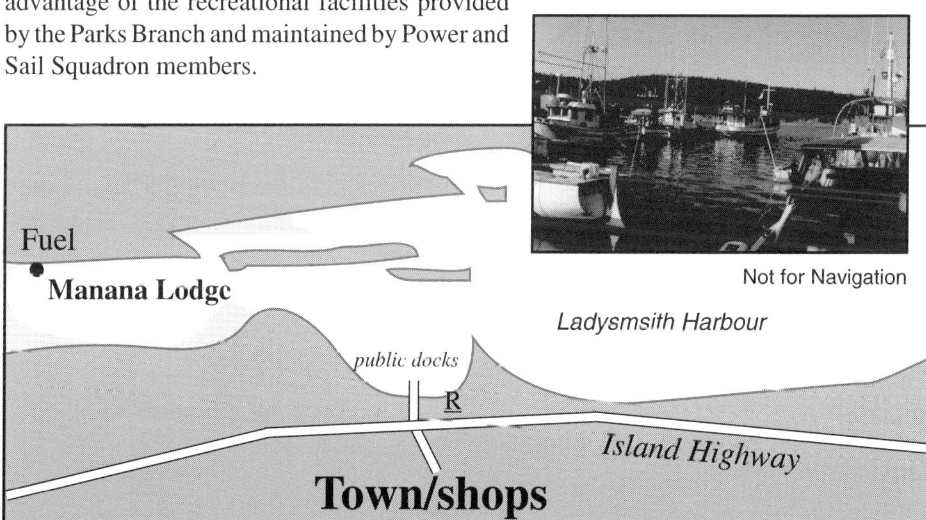

Chemainus Transport Canada dock Manager • Float length 175 m *Good stop for visiting uptown Chemainus, internationally famous for its artistic murals.*	**Crofton,** Vancouver Island Fisheries & Oceans dock Phone (250) 246-4512 Manager • Float length 158 m Launch ramp • Breakwater • Garbage • Water • Lights • Power • Public phone.

Ladysmith Harbour

Manana Lodge & Marina

Chart 3313, 3475, 3443, 3463 VHF 68

Don and Gail Kanelakos,
Jim and Ruth Bangay
4760 Brenton-Page Road, RR #1
Ladysmith, B.C, V0R 2E0
Phone: (250) 245-2312 Fax: 245-7546

Marina services:

Fuel: Gas. Diesel. Propane. Service available. Marine supplies.
Fishing gear, licences.
Guest moorage, 1400 feet dock space.
Water at dock.
Power: 15amps. 30 amps to be added.
Laundry, showers, washrooms.

Customer services:

Restaurant. Fine dining. Licenced.
Also patio service.
Gift shop.
Public pay phones ashore.
Daily float plane service accessible and stops at marina for pick-up.

Entertainment.

Cycles free with room rental. Canoes, rowboats. Car rentals. Road access walking or cycling.
Scuba diving arrangements and charters–ask marina for details.

Adjacent facilities:

Excellent bed and breakfast accommodations at lodge. Make reservations ahead in summer months.
Short boat trip to Ladysmith–stores–stop at government dock opposite/south just inside entrance to Ladysmith harbour.

Ladysmith,

Vancouver Island
Fisheries & Oceans dock
Phone (250) 245-7511

Manager • Float length 213 m
Launch ramp • Breakwater • Grid • Garbage •
Waste oil disposal •
Water • Lights • Power • Public phone •
Near uptown restaurants, shops.

Manana Lodge and marina offers excellent overnight sheltered moorage, fuel, restaurant and many amenities.

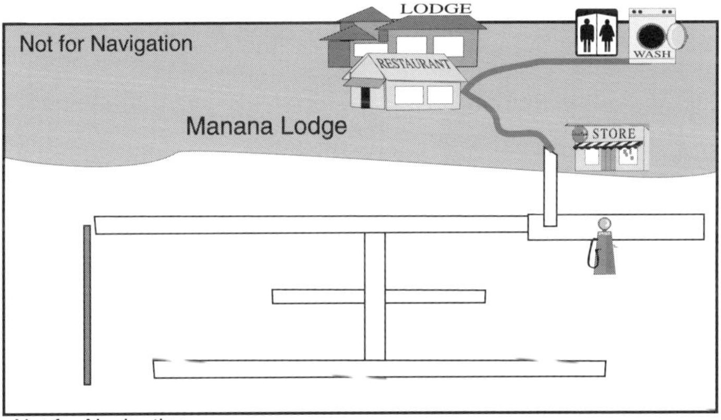

Not for Navigation

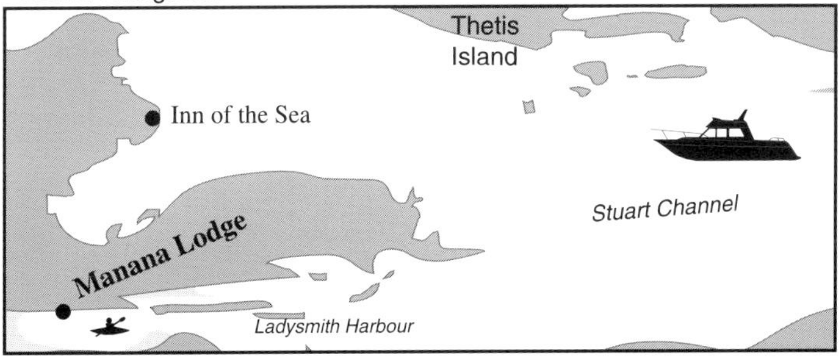

Inn of the Sea Resort

Carol Radloff
3600 Yellow Point Road,
Ladysmith, BC. VOR 2E0
Phone (250) 245-1011 Fax (250) 245-1023
Chart 3313, 3443

Customer services:

Restaurant. Lounge. Breakfast, lunch, dinner. Licenced.
Luxurious shore accommodations. Heated outdoor pool.
Road access walking or cycling. Nearby walking trails.
Public pay phones ashore.

Hazard: Reef extends out from the point. Give wide berth and approach from the south-east. Consult chart.

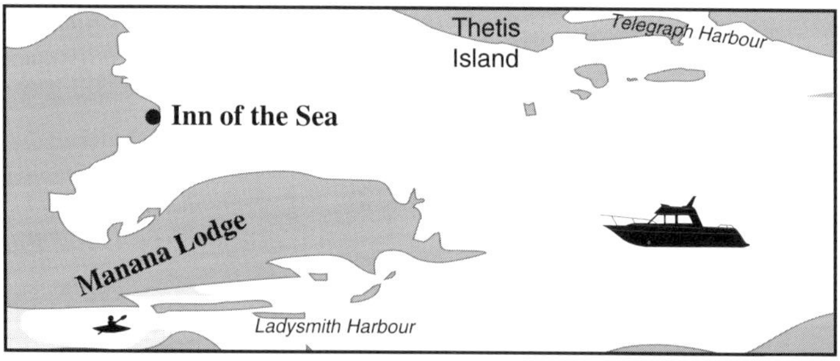

Not for Navigation

Marina services:

Guest, limited moorage.
Reservations advised.
No water at dock.
Power at docks: 15 amp.
Mooring buoys.
Laundry, showers, washrooms at lodge.

Entertainment.

Tennis, canoeing, swimming, jaccuzi.

Adjacent facilities:

Bed & breakfast accommodations at lodge.

The alluring Gulf Islands

In the Gulf Islands you will probably linger longer than intended when you discover such treasures as Ganges and its Saturday morning market, its fine restaurants, arts and crafts, clothing stores, food stores and the famous Mouat's Trading Store, or Port Browning with its nearby weekly farmer's market. Or simply the serenity of tying to a mooring buoy at Montague Harbour or the dock at Wallace Island as you watch the sun go down. You may never want to leave Telegraph Harbour when you walk up to the restaurant for a 50s style milkshake or the Thetis Island Marina pub for a hearty burger, or watch the spinning of wools and other fibres at the nearby Alpaca and arts and craft farm.

The charming Gulf Islands hospitality may do it for you, or the nearby Vancouver Island stop at Chemainus with its great historic murals, or an evening dining at the Manana Lodge restaurant on the waterfront in Ladysmith Harbour.

Pub atmosphere is sought by some and if you missed the one at Thetis stop at the Vesuvius pub or Nanaimo's Dinghy Dock Pub any of which will bring you back time and again. Cross the Nanaimo harbour to the Boat Basin with its Lighthouse Pub and Bistro, or spend the night farther up at Stones Marina with its Muddy Waters Pub and restaurant.

There's a good deal to see and do at Nanaimo and if you arrive at the time of the Bathtub Race in July you will want to be on hand to see the spectacular send-off and other activities that take place in the town. You may just want to enjoy a long walk on Newcastle Island, one of the coast's prime marine parks. Marine parks are featured in ***Anchorages and Marine Parks***, the companion book to this one (*published 1998*).

Wallace Island marine park offers good walking trails as well as beautiful views of Trincomali Passage and the adjacent Gulf Islands. It's a popular anchorage with a small dinghy dock. It has extensive walking trails. For more on marine parks see **Anchorages and Marine Parks.**

When entering a sheltered moorage area take care. Check your wake well before you arrive at the moorage and slow down a good distance off. Check again and if it is still creating a large wash ease off further on the throttle. Your wash should not be big enough to be felt in the marina. It will follow you in, possibly overtake you and it can damage boats at the floats. Anyone who creates a wash can be held liable for any damage it causes.

Anchorages and trails such as those found at Wallace Island are features of the Gulf Islands. The joy of simply being there during the course of any summer is difficult to match. And there is much history to the area which can be uncovered by delving into the origin of some place names.

Photographs show a pathway on Wallace Island, top, a signpost at the trail head, left, and the anchorage at Conover Cove, below.

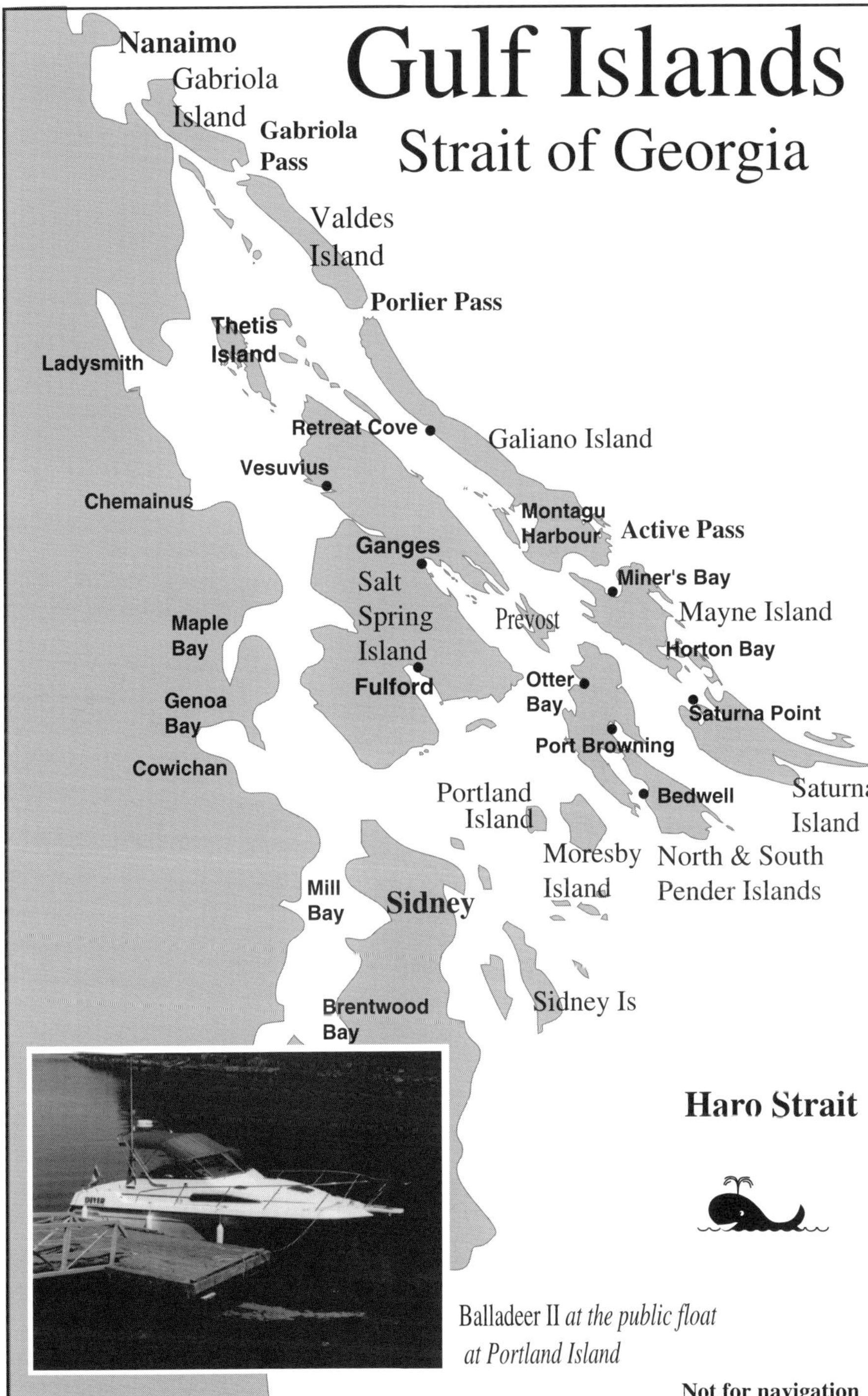

Balladeer II *at the public float at Portland Island*

Salt Spring Island

Top: The church at Fulford Harbour was built using materials carried to the island in canoes. Above: Looking up Fulford Harbour from across Portland Island.

The islands in the lower Strait of Georgia known as The Gulf Islands offer many anchoring and mooring alternatives. The diagram on the previous page shows the main archipelago which makes up the group of islands that are most popular for overnight use. Note the cautions in the following text but always be mindful of weather conditions and forecasts.

Salt Spring Island is the largest of the Gulf Islands and has most facilities and amenities similar to mainland and Vancouver Island centres.

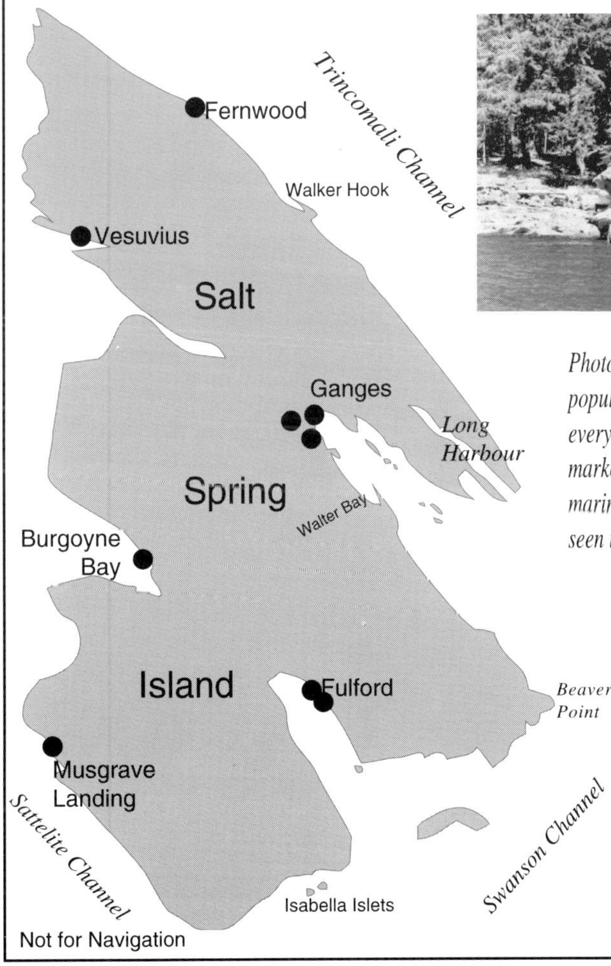

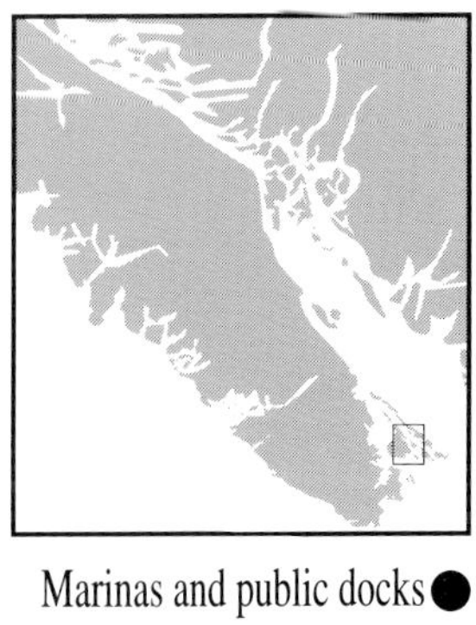

Photograph top: A view over Ganges, the popular Gulf Islands destination where every Saturday morning the farmers market attracts a large crowd of visiting mariners. Above: Sandstone cliffs can be seen in many parts of the Gulf Islands.

Fulford Harbour

Chart 3313, 3478, 3441/2, 3462

Fulford Marina

Dock manager
#1–2810 Fulford-Ganges Rd.,
Salt Spring Island,
BC. V8K 1Z2 VHF 68
Phone: (250) 653-4467 Fax: 653-4457
email: *fulfordmarina@saltspring.com*
Web: *www.saltspring.com/fulfordmarina*

Marina services:

Moorage: Guest and permanent.
Large breakwater float.
Fuel: Gas, diesel, oil.
Water at dock. All slips.
Power at docks: Multiple 30, 20 amp.

Customer services:

Public phone ashore.
Showers, washrooms.

Entertainment.

Island tranquility. Eagles, herons, otters and seals. Tennis courts, waterfront gazebos and barbecue area. Walking, beaches.

Nearby facilities:

BC Ferries to Sidney. Government docks. Places of interest: Historic churches. Fulford settlement–arts and crafts, groceries, restaurants, hotel. Walk to Drummond Park. Nearby private museum of Indian Art–Bob Akerman collection and his wife's classic doll collection. Five minutes walk to pub/ dining room, grocery store, coffee shop, crafts, parks, ferry. 10 minutes to lake.

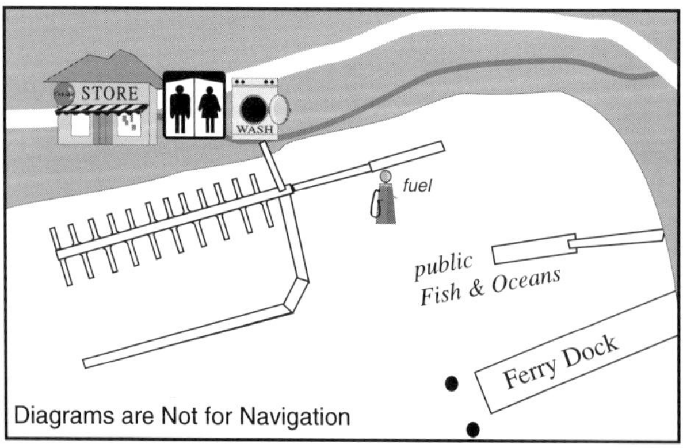

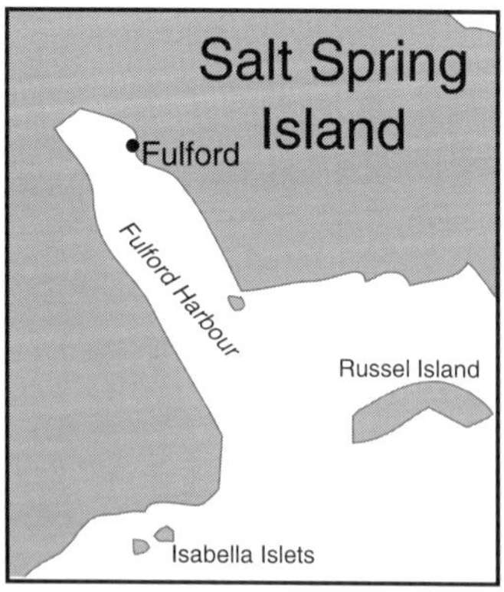

The restaurant at the ferry landing at Fulford, known for many years as Rodrigo's.

An old fuel pump at what was once a busy garden shop in Fulford Harbour

Public docks:
Fulford Harbour, inner dock
Fisheries & Oceans dock
Phone (250) 537-5711
Chart 3313, 3478, 3441, 3442, 3462
Manager – Reid Collins.
Phone (250) 653-4728
• Float length 36 m
• Garbage • Lights • Power • Public phone ashore
Adjacent ferry to Sidney, restaurants, shops.

Fulford Harbour, Outside breakwater.
Transport Canada dock
Chart 3313, 3478, 3441, 3442, 3462
Manager • Float length 16 m
Public phone ashore •
Not recommended for overnight moorage.

Fulford to Ganges

Salt Spring Island is a busy destination for holiday makers every summer. The active hub of the island at Ganges sees throngs of visitors and locals every Saturday to the open air market and the grocery and general merchandise stores. The restaurants do a roaring trade and the streets are abuzz with motorists, cyclists and pedestrians intent on going nowhere in particular in no great hurry. The island, like a diamond in the making, has many facets. Where Ganges is alive and active, other areas of the island are sleepy and quaint or just plain laid-back like granddad dozing in the afternoon.

At the south end of the island Fulford Harbour is one of those places you may never visit by boat if you aren't prompted to go there for some specific reason. For years I by-passed Fulford Harbour en route between Vancouver and Sidney, thinking there was no point running up a blind inlet in a slow boat. I was wrong. I had missed out on what is a favourite cruising destinations for the many who have discovered its charms.

Fulford Harbour has some of the Gulf Islands' most fascinating features and history. The oldest church on the island is located not far from the marina, the historic Fulford Inn nearby and the village at the ferry terminal give this quaint place a character all of its own, making it quite different from other settlements in the Gulf Islands. Fulford is the start of the highway across Salt Spring from the ferry terminal that carries traffic out of Sidney, and gives travellers and residents access to and from the island's other settlements, Ganges being the major centre.

Summertime sees thousands of visitors passing through the gateway village of Fulford, many stopping at the busy restaurant at the ferry terminal, or historic Patterson's general store and gas station next door. Throngs of passengers and islanders picking up foot-passengers can be seen almost constantly awaiting ferry arrivals and departures. They sit at tables outside the stores, on the roadway or in their cars. In summer tourists come by bicycle, camper, car and motorcycle and bring with them tents for camping at any one of the many beautiful campsites on the island. Some travellers arrive at Fulford and leave from another ferry terminal, usually Long Harbour, or the other way around.

At the Fulford settlement there is a surprising number and variety of enterprises and services. At the post office you can buy stamps as well as fishing licences. Not far you can walk into the local craft shop or cross the road and browse in the very well-stocked nursery, eat at the adjacent restaurant–inside or out on their patio, or check out the goods at the pottery studio, or the knick knack store nearby.

No matter what interests you at Fulford or what other people are up to, waiting for a ferry or just passing through, stopping there even for a brief moment is certain to captivate one by its charm and tranquillity.

Arriving by boat up the harbour, a wide and sometimes windy inlet, one can stop briefly at a small government dock outside the breakwater, or enter behind the breakwater and possibly find a spot at the government dock inside. But moorage is readily available at the Fulford Marina most times of the year.

In winter time some fairly strong winds can assail Fulford, blowing up the harbour from the south east, so monitor the weather channel before planning a trip there out of season. If you do go into the marina and anticipate a storm, make sure you tie up well inside rather than on the outer floats. The docks are subject to a fair amount of movement in a storm. A large log boom has been used as a breakwater during winter. Overnight the Fulford ferry remains at her berth, serving as additional protection to inner sections of the marina.

One of the best attractions at Fulford Harbour is the pleasant surroundings and the ability to go for a quiet stroll ashore. From the marina a path follows the immediately adjacent shoreline and gives access to a tranquil spot under a picturesque gazebo overlooking the marina and the harbour entrance. Following the pathway leads to the island highway which passes tree sheltered tennis courts (take your tennis gear with you and check with the marina) and leads down to the Fulford Inn. Along the way you pass the quaint old St. Paul's church which has an interesting history. It was built back in 1880 by the local population using materials recovered from the remnants of another old church in the Gulf Islands. A stained glass window, the door and a bell were carried from Cowichan to Salt Spring Island by Indian canoe and then hauled overland by an ox-drawn cart on skids.

Once at the Fulford Inn you can stop for lunch or make dinner reservations, or just drop in at most meal times. It's a busy place at times with some Christmas and other holiday meals booked to capacity. The Inn, as old and historic as it may appear, however, is relatively new, being the fourth building on its site, the previous three having succumbed to fire. Not far from the Inn is the community hall and another old church, St. Mary's, which was built back in 1894 and, like St. Paul's, continues holding services to this day.

We have visited Fulford in the winter and found some facilities operating at reduced hours. The Inn and restaurants are excellent places to meet the local islanders. Some of these are internationally famous figures. Salt Spring Island is home to many artists and their work can be seen at local craft stores and galleries on the island. Beyond the Inn on the road to Isabella Point is cosy little Drummond Park where you can sit on the benches or swings and watch the sea birds and other animals that visit the beach. We have seen swans, geese and sheep wandering on the beach and roads in the area. The park has a playground for children and toilets as well as a large rock bearing a native petroglyph. This area was once a popular hunting ground for aboriginal people and middens are apparent in places along the shore.

In the opposite direction, down the harbour shoreline beyond the ferry dock takes you along a winding road to a park area and Indian reserve. The shady pathways that meander all the way down to the harbour entrance make an excellent outing in summertime or sunny days in the fall or spring. Stop along the way and watch the occasional harbour traffic or just sit and enjoy the ambience of the area. Look for deer, eagles, herons, woodpeckers, hummingbirds and many other types of indigenous birds.

After a long walk or time spent exploring the settlement of Fulford there is little to match the joy of sitting in your boat and watching the activity at the dock or exchanging communications with neighbours while barbecuing or preparing and having a sunset dinner.

At Ganges the Saturday morning market is a great attraction. Opposite: Fuel dock at Fulford Marina seen from the government dock just across the way.

Ganges Harbour can be very crowded in summer. If you anchor in the harbour be mindful of potential winds, which have been known to cause great consternation at times. Make reservations in advance for a weekend at Ganges. Book into one of the marinas and plan to visit uptown bakeries and restaurants. There are some fine establishments to try. Walk around the village and take in the local atmosphere. Stop in at the various stores and chat to the staff. View the local arts and crafts and try some of the local treats, such as ice cream.

The Saturday morning market offers a wide variety of interest. There are booths and stalls featuring clothing, collectibles, trinkets, food, fruit and vegetables, stained glass, ornaments, toys, wood carving and other local crafts. On a sunny day this is also a popular photographic opportunity.

Rent a car or motor scooters and go exploring around the island. There are pubs, restaurants, golf courses, and some breathtaking views along the way.

Ganges

Ganges Marina

Suzanne Guiness
161 Lower Ganges Road,
Ganges, BC V8K 2T2
Phone: (250) 537-5242 Fax (250) 537-4322
email: *guinness@saltspring.com*

Marina services:

Moorage: Guest moorage only. Reservations taken.
Fuel: Gas, diesel, oils–Shell. Other services available.
Water at dock. All slips.
Power at docks: 50, 30, 15 amps.

Customer services:

Public phone ashore. Marina store has some supplies. Ice.
Laundry, showers, washrooms.
Complimentary bikes and baby strollers. Also coffee, muffins and juice mornings and *hors d'oeuvres* on long weekends.

Entertainment:

Ganges Saturday public market.
Crafts and art galleries.
Scooter and car rentals at marina. Walking and cycling on island roads, some nearby waterfront and beachfront access.

Nearby:

Government docks. Shopping centre. Propane. Close to all Ganges facilities. Fresh produce. Bakery. Thrifty Foods (537-1522)–groceries. Restaurants. Anchorages and places of interest. Churches. Arts and crafts. Hotels, bed and breakfast. Bistros. Pubs.

Outdoor restaurant patio at Mouat's in Ganges.

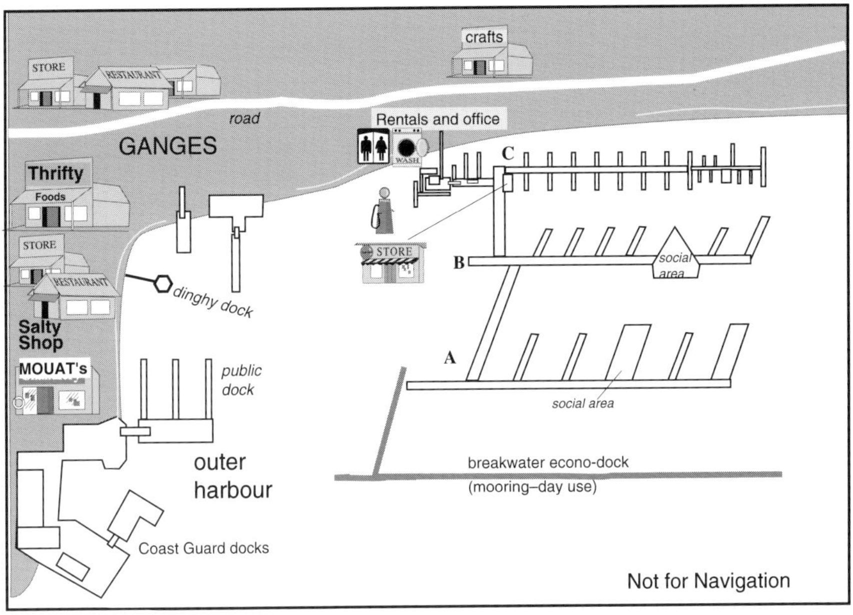

Photograph opposite and diagram above show Ganges Marina occupying a large portion of the bay. The public dock, abutting Mouat's large store building, is located to the right of the marina and just beyond the Coast Guard dock.

Ganges is the business centre and hub of the Gulf Islands. It is not only located on the largest of the Gulf Islands but also it has the largest population of all communities in the archipelago. Salt Spring Island residents and visitors by boat or by road via the BC Ferries system use Ganges as a shopping and cultural centre.

Activities on Salt Spring Island as well as arts and crafts attract many visitors each summer. The morning farmers market held on the waterfront every Saturday has become a colourful attraction. The work of local artists can be seen and bought in the several art shops and galleries in Ganges. Shopping at Mouat's historic store provides opportunity to stock up on the items you need for your boating comfort, safety and convenience. The many other speciality stores in the town will provide hours of pleasurable shopping or window shopping. And the restaurants are of a variety that will enable you to select from a wide range of menus. Hastings House, one of the top restaurants in Canada, is located in Ganges.

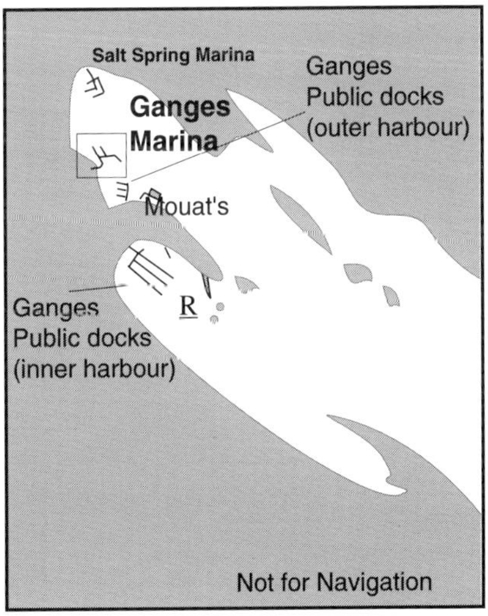

Charts 3313, 3462/3 , 3478

Salt Spring Marina

Dick Durante. Lesley Cheeseman (mgr)
120 Upper Ganges Rd.,
Ganges, BC. V8K 2S2
Phone: (250) 537-5810 Pub: 537-5559
Chart 3313, 3462/3, 3478 VHF 68

Marina services:

Moorage: Guest and permanent.
Haulouts, towing and other services.
Water at docks.
Power at docks: Multiple 30, 15 amp.

Customer services:

Restaurant and pub. Public phone ashore.
Marina store has chandlery, fishing licences and tackle. Marine service.*
Showers, washrooms. Grocery store, souvenirs, coffee, ice, bait, fishing gear.
Car rentals at Ganges Marina.

Entertainment.

Ganges Saturday public market.
Arts. Galleries.
Scooter, kayak and boat rentals.
Walking on island roads, some nearby waterfront and beachfront access.

Adjacent and nearby facilities:
Moby's Pub. Shopping centre.
Anchorages and places of interest:
Churches, arts and crafts, groceries, restaurants, hotels, bed and breakfast.

* Harbour's End Marine & Equipment Ltd

Barry Green and Ross Walker
122 Upper Ganges Rd.,
P.O. Box 1440 Ganges, BC. V8K 2S2.
Phone(250) 537-4202
Marine repairs, chandlery, service.

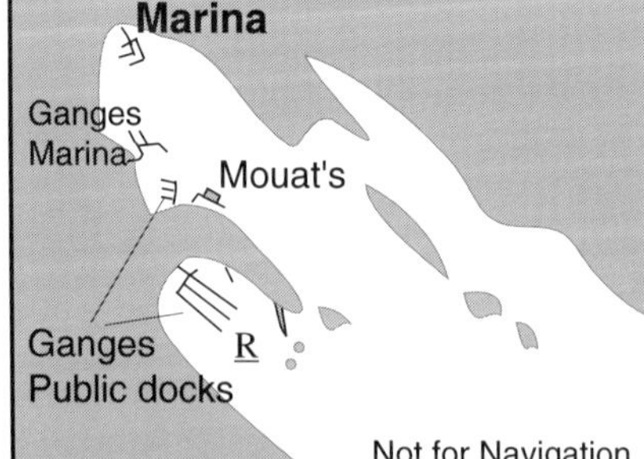

Hastings House restaurant will take reservations and collect you at your boat for a meal that will cost you a lot but provide you a memorable gourmet experience that will be hard to match.

Boat services include temporary stopping if space permits, at the government dock on the north side of the town or in the harbour. Private marinas Ganges Marina and Salt Spring Marina provide overnight and extended mooring with power and electricity at the docks and laundry, showers and toilets on shore.

Salt Spring Marina offers also a large restaurant and pub at the head of the dock and activities such as boat, bicycle and scooter rentals, as well as boat repairs and service. Harbour's End Marine and Equipment is an authorized dealership for several brands of engines and other equipment and is located right at the marina just opposite Moby's pub and restaurant. Ganges Marina nearby sells fuel.

Market at Ganges attracts crowds on Saturday mornings throughout the cruising season. Merchants from local stores and local artisans conduct business from these outdoor stalls, offering a wide range of crafts and merchandise along with some tasty snacks and refreshments.

Not for Navigation

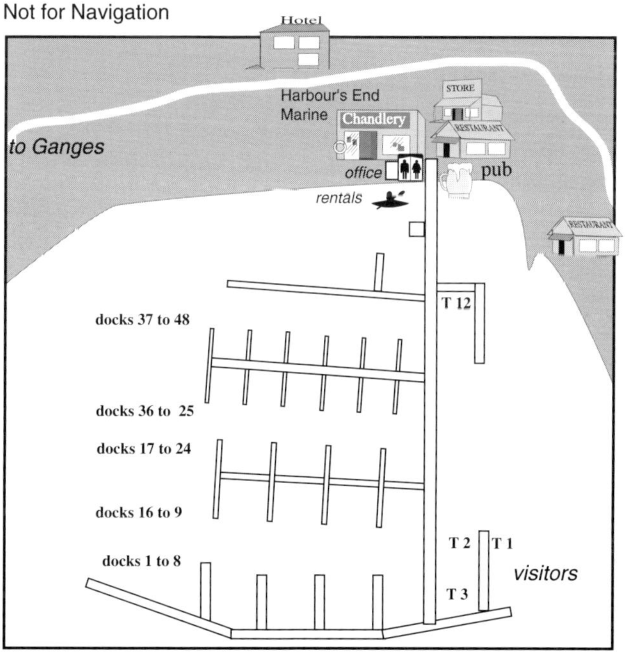

A note of caution: Walking from the marinas into town or to other areas can be pleasant. However, caution should be exercised due to the sometimes heavy traffic on the roads and limited pedestrian sidewalks. Walk facing the traffic where there are no sidewalks.
A waterfront boardwalk has been under construction for some years but the sections have not yet been connected due to unavailability of land in one area.

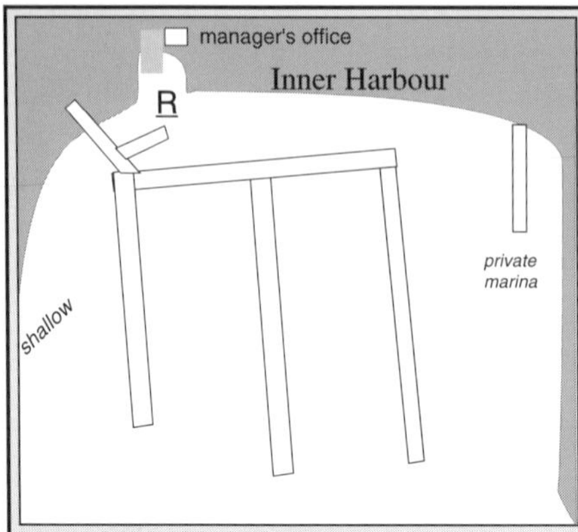

Ganges, Salt Spring Island Public docks.

Ganges Boat Harbour (Inner Harbour)

Fisheries & Oceans dock

Phone (250) 537-5711

Manager • Float length 326 m
Launch ramp • Breakwater • Garbage • Waste oil disposal • Water • Lights • Power • Public phone • Washroom • Showers.
Adjacent to restaurants, shops.

(Outer dock and Coast Guard)

Transport Canada dock
Float length 24 m
Aircraft Float • Water • Lights • Power • Public phone ashore •
Adjacent restaurants, shops.
Fuel dock adjacent.
Marinas, fuel, services nearby.

Ganges, (Outer Harbour)

Fisheries & Oceans dock
Manager–*check in at office in Inner Harbour for stays over two hours* • Float length 41 m • Garbage • Water • Lights • Public phone ashore •
Adjacent to restaurants, shops.
Phone (250) 537-5711–for all public docks at Ganges.

Charts 3313, 3478, 3442, 3462

The outer harbour public docks and the Ganges waterfront. Mouat's, Thrifty's, The Salty Shop and other stores are located here.

Fernwood,

Walker Hook, Salt Spring Island
Transport Canada dock
Chart 3313, 3462, 3442
Float length 12 m

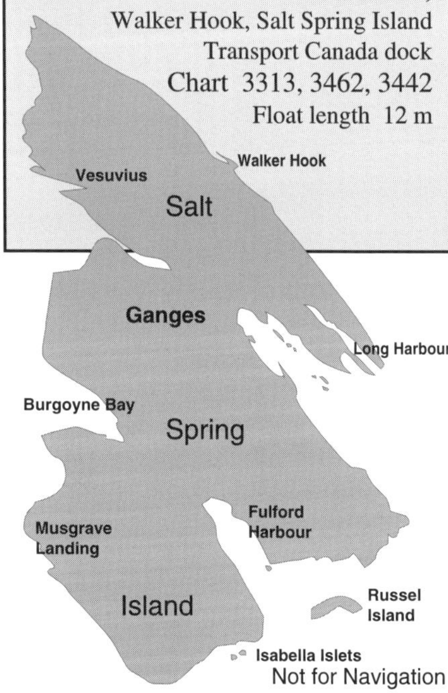

Burgoyne Bay
Fisheries & Oceans dock
Phone (250) 537-5711 Chart 3313, 3478 • Manager • Float length 10 m

Musgrave Landing, Salt Spring Island Fisheries & Oceans dock Chart 3313, 3441/2 Float length 12 m

Top: Inner Harbour behind the breakwater on the south side of Ganges. Note the small private marina to the right.
Above: View of Ganges Outer Harbour small dock right of Ganges Marina (centre). Inner Harbour far right. Salt Spring Marina far left.
Left: Boats at the outer harbour public dock.

Top: Pender Canal, the narrow passage between the Penders was man made to provide an alternative route north to south. At high tide the clearance is 26 feet and at low tide the depth of water is seven feet.

Centre: Port Washington general store which closed down in the past decade after serving passing vessels since the early 1900s is now an art gallery. The floats will accommodate several small to medium sized boats but are exposed to washes from passing ferries and some wind conditions.

Bottom: One of the fuel stops in the Gulf Islands is at Bedwell Bay.

The Pender Islands

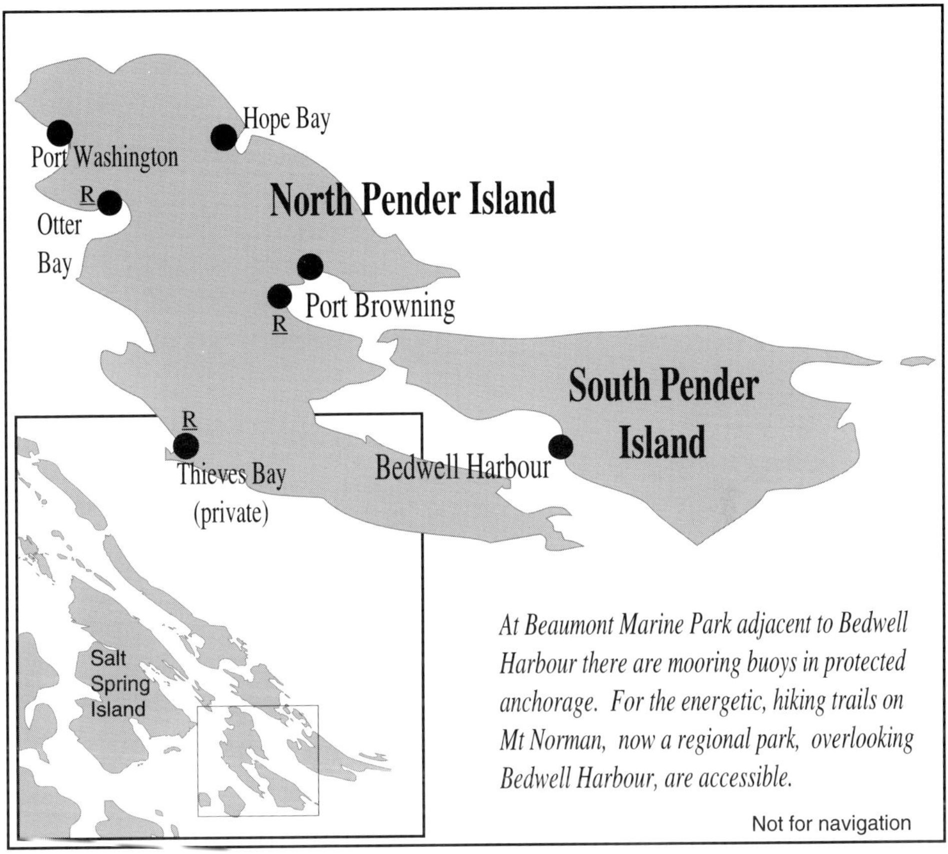

At Beaumont Marine Park adjacent to Bedwell Harbour there are mooring buoys in protected anchorage. For the energetic, hiking trails on Mt Norman, now a regional park, overlooking Bedwell Harbour, are accessible.

If there were no islands in the Strait of Georgia other than the Pender Islands we would still be talking about the outstanding cruising on the west coast of Canada. The Pender Islands are blessed with beautiful, unspoilt and charming features and are central to the major group of Gulf Islands, providing moorage, anchorage and safe stops for all boat operators. The main harbour of the Penders is Bedwell: resort, marina and customs stop. It is a major anchorage with an adjacent marine park which ranks highly in popularity among boaters and campers. Up the harbour is one of the most picturesque passages on the entire coast, Pender Canal, which separates North from South Pender and provides access to Port Browning. That marina facilities include pub and restaurant, store and swimming pool as well as laundry, showers and close access to the island's main but small business centre. Here one can shop for groceries, arts and crafts, liquor, real estate, pharmacy items and more. A popular stop is the ice cream and bakery store as well as the Saturday Farmers' Market at the shopping centre's property. In late summer look for ripe blackberries on the bushes that line the roadways.

Pender Canal is a narrow passage which curves its way under a low bridge that

The floats at Hope Bay were more extensive at one time, serving as an alternate landing for the Otter Bay ferry during strong westerly blows. This is also an exposed dock and best used for short visits. Gone is the Hope Bay store, an early establishment on the Penders which burned down in the winter of 1997/8. Opposite: A timeless landmark, the restaurant and pub at Bedwell Harbour.

connects the two Pender Islands. At high tide boats will clear the bridge if they are no taller than 26 feet. At low tides shallow draft vessels have no trouble but a seven foot draft could be a problem. Travel slowly through the canal and be prepared to possibly meet a boat coming from the opposite direction midway through.

A major, but little known attraction on North Pender Island is Hope Bay named after a former postmaster and local land owner, Rutherford Hope. The docks at Hope Bay are minimal and exposed to wind at times making them not the most secure for overnight moorage. There is no power, no water and no services. But on land is the remains of the historic Hope Bay Store which in recent years closed and reopened under different owners before suffering a devastating fire in the winter of 1997/98. It served for many years as post office, supply store and provision centre for the islanders living at that end of Pender as well as for passing vessels or those calling from neighbouring areas, such as lighthouse keepers and islanders of nearby islands. At one point it was also a ferry landing.

The docks at Hope Bay are not too extensive but can accommodate several medium sized craft on the outside of the main float for short stays, and another several on the inside if the docks are clear or if you are prepared to raft alongside other boats.

General stores seem to have enjoyed a heyday in the islands. On North Pender Island is the Port Washington General Store on the west side of the island in Grimmer Bay. It

served the local people from 1910 until the late 1980s and in 1998 was being operated as an art gallery. Washington Grimmer was responsible not only for the names of the settlement and the bay, but also for the mail on Pender in the early years and was involved in bringing the first educator to the island. There is a government float at Port Washington which will accommodate about four medium sized boats at each of two fingers. Some wash comes into the bay from passing ferries and other vessels or from weather conditions from the west.

En route from Port Washington by foot or bicycle to other parts of the island you will find local arts and crafts for sale at private and semi-private homes. By boat or by road, a short distance south is the Pender Islands ferry terminal at Otter Bay. Here is another of the Gulf Islands' major ports for recreational boating.

Otter Bay has a fully operational marina with good docks, power, water, laundry, showers and a store which carries a limited amount of stock; local crafts, paintings, clothing, books, some fresh produce, canned goods and frozen foods among other items. The marina office sells fishing licences and can provide information on local island services such as mechanical, towing, bed & breakfast and ferries schedules. It is a convenient spot to leave your boat between weekends and ferry back and forth between Pender and Vancouver Island or the mainland.

The facilities at Otter Bay marina are exemplary. New, clean washrooms, showers and laundry, large sun deck on top of the breakwater and a swimming pool are available to the mooring visitor. A large lawn is a playing field for the children and several picnic tables have been placed along the waterfront. If you want a catered meal go to the golf club nearby for breakfast or take in one of the local restaurants whose dinners are legendary.

There is no fuel available at Otter Bay. The nearest fuel docks are at Bedwell Harbour, Ganges, Active Pass, Montague Harbour, Saturna, Fulford or Sidney.

Bedwell Harbour

Bedwell Harbour Resort

Bill James

RR #1 South Pender Island,
BC., V0N 2M0
Phone: (250) 629-3212
Fax: (250) 629-6777
email: *bedwell @islandnet.com*
Web: *www.islandnet.com/~bedwell/*

Customs service.

Customs. Phone (250) 629-3363.
Officials on duty office and other hours.

Marina services:

Fuel: Gas, diesel, ice. Service.
Moorage. Large marina with mostly transient moorage.
Water at dock–limited supply.
Power at docks: 15, 30 amp.
Laundry, showers, washrooms.
Tackle shop on dock.

Customer services:

Restaurant. Pub. Breakfast, lunch, dinner. Licenced. Also patio service. Swimming pool. Lodging. General store: Fishing gear, licences, charts, bait. Fresh baked goods. Gifts. Groceries.
Tennis raquets available. Bicycle rentals. Road access walking or cycling. Nearby hiking trails.
Public pay phone ashore.

Entertainment.

Live music weekends, summer. meals at poolside. Tennis courts.
Fireworks July 1.

Adjacent facilities:

Play area. Pet area.Beaumont Marine Park. Camping, hiking. Mooring buoys. Nearby Mt. Norman is accessible by trail.

This is one of the main American entrances to Canadian waters in British Columbia. It is a busy customs port throughout the summer months June through September. It is essential that visiting vessels clear customs at an official entry port before stopping elsewhere in Canada. And Bedwell is one of

Opposite and right: Bedwell Harbour Island Resort and marina.
Below l & r: The marine pub and patio with general store to the left. The pool overlooks the marina.

the most used entry ports for vessels arriving, including returning Canadians, out of the San Juan Islands or Puget Sound. The facilities at Bedwell are as comprehensive as you will find anywhere in British Columbian waters. The docks are extensive and power and water are available, the latter not for washing boats due to the relative short supply from its on-island source. A fuel dock serves gas and diesel and other marine products and the marina, as part of the shoreside resort, offers access to the swimming pool and other amenities included with one's moorage. Showers, laundry, a general store, restaurant, snack kiosks, and a marine bar and bistro make up most of other facilities. There is a hotel which is popular for visitors arriving by boat or by road via ferries from Sidney or the mainland, as well as town-houses which are home to their owners who mostly use them at weekends.

Chart 3313, 3477, 3441 VHF 68

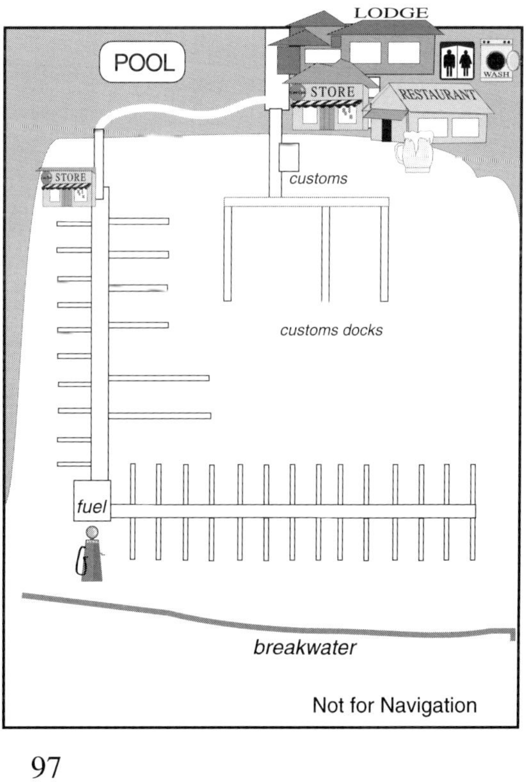

Port Browning

Chart 3313, 3477, 3441/2 VHF 68

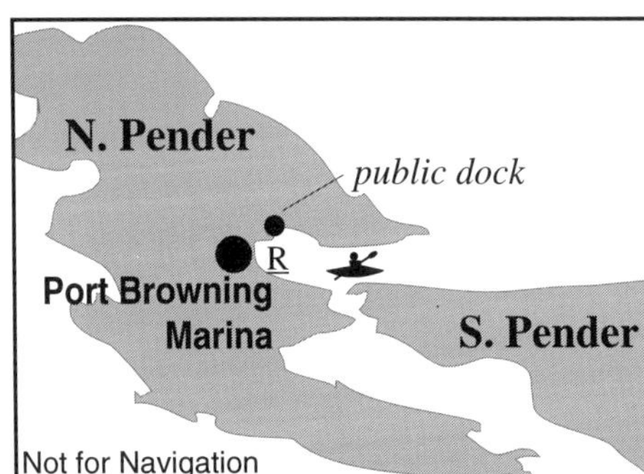

Port Browning Marina Resort

Lou Henshaw
General Delivery,
North Pender Island.
BC, V0N 2M0
Phone: (250) 629-3493

Caution: Narrow, shallow Pender Canal if travelling from Bedwell Harbour. Check chart and tides.

Customs service.
Check in at Bedwell Harbour.

Marina services:

General and marine store. Ice, groceries, charts. Fishing tackle, licences, ice, bait, gifts. Moorage. Large marina with permanent and transient moorage.
Water at dock.
Power at docks: 15 amp.

Laundry, showers, washrooms.
Reservations advised summertime.

Customer services:

Pub. Restaurant. Breakfast, lunch, dinner. Licenced. Also patio service.
Swimming pool. Cold beer and wine store at marina.
General store: Fishing gear, licences, charts, bait. Fresh baked goods. Gifts. Groceries.
Road access walking or cycling. Take care

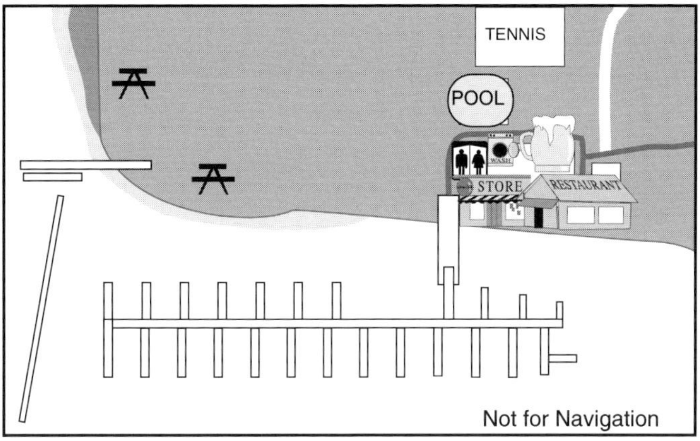

Docks and facilities at Port Browning include pub, store, restaurant, pool and tennis courts. A large expanse of lawn is used frequently in summer for yacht club or group gatherings and regular weekend lamb or pork barbecues.

Farmers' market at nearby shopping complex. (Below)

walking the narrow island road.
Good beach access.
Public pay phones ashore.

Entertainment.

Swimming pool, large lawn, camping.
Tennis, golf, nearby.
Saturday morning early farmers' market at nearby shopping centre.

Adjacent facilities:

Shopping centre nearby: wide range of services, general store, restaurant, liquor, bank, art, gifts, dairy products, bakery, hairdresser. **Small craft launch ramp.**

Browning Harbour (public)
North Pender Island
Fisheries & Oceans dock
Phone (250) 629-3423
Chart 3313
Manager • Float length 27 m
Store, shopping complex nearby.
Port Browning Marina is located almost directly opposite.

Otter Bay

Otter Bay Marina

Kay and Chuck Spence
General Delivery,
North Pender Island.
BC., V0N 2M0
Phone: (250) 629-3579

Hazard: Keep clear and to starboard of green spar U 57.

Marina services:

Launch ramp (small craft).
Moorage. Large marina with permanent and transient moorage.
Water at dock. Use sparingly please.
Power at docks: 30, 15 amp.
Laundry, showers, washrooms.
Reservations advised summertime.

Customer services:

Store. Gifts, bait, ice, limited groceries. Charts. Fishing tackle, licences.

Chart 3313, 3441/2 VHF 68

Garbage disposal.
Road access walking or cycling. Take care walking the narrow island road.
Public pay phones ashore.
Cabin sleeps 4–5 people.

Entertainment.

Eagles, herons, otters and seals.
Bicycle, boat and kayak rentals.

Adjacent facilities:

BC Ferries dock.
Picnic tables, barbecue area.
Gazebo.
Restaurants nearby.
New deck on breakwater at marina.
Golf course–10 minutes walk.

Crew off anchored boats welcome ashore for a small fee.

Otter Bay Marina is a very popular resort. Week ends in mid summer and on holidays are booked well in advance. For many years Otter Bay operated with minimal dock space and limited service. When Chuck and Kay Spence took over running the facility they set about enlarging the docks, replacing the buildings on shore and adding amenities to make a visit to the bay a more enjoyable experience. They have achieved this and more with the provision of a swimming pool, coffee bar and store. The pool has hours set aside for adult use only which is popular among those needing to relax undisturbed while they bathe during a hot day.

Otter Bay was not always the laid back cruising destination it is today. At one time it flourished as a fishing centre featuring such establishments as cannery, saltery and reduction plant. But in 1963 it took on a new life as a marina and by 1972 it had moved to the ownership of Bob Melville who many of today's cruising yachtsmen got to know quite well. After nearly twenty years Bob and Karen Melville moved to the BC Interior when David Bromley bought the marina and the Spences took over managing it. The Melvilles left a legacy at Otter Bay in the way of Karen's flower and vegetable garden as well as the fruit trees that can be found on the property.

Chuck and Kay Spence have created their own legacy with the additions they have made to the property. Their arts and crafts selection, the covered barbecue area and their friendly welcome to moored guests has become their trademark.

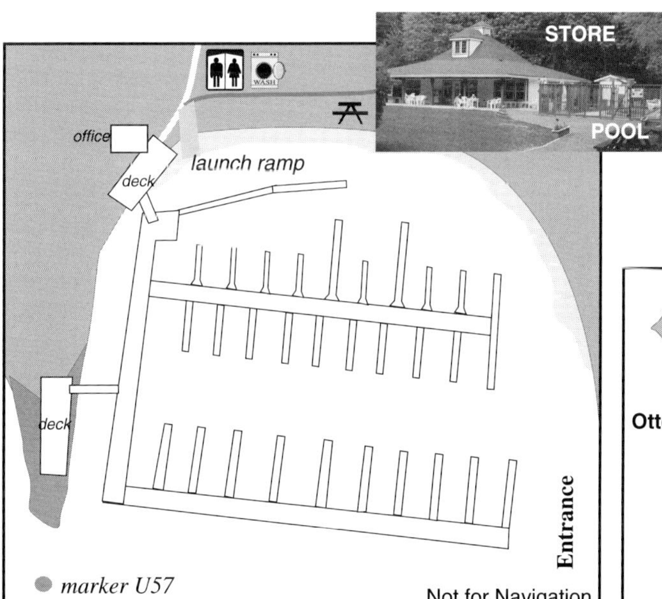

Above: Summer months are busy so make reservations ahead of your arrival. Inset, left: The store and adjacent pool.

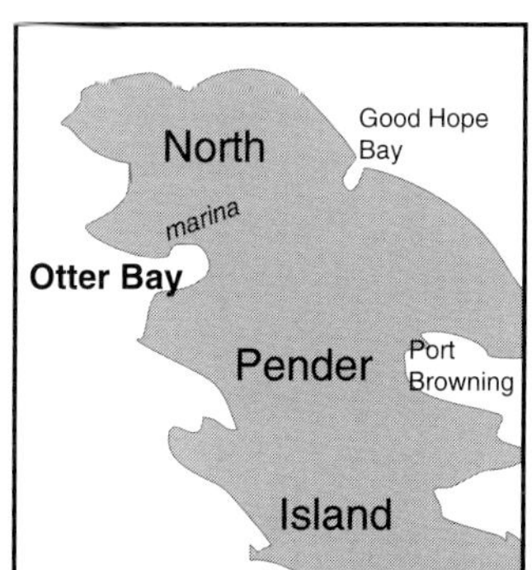

Aerial photo page 92

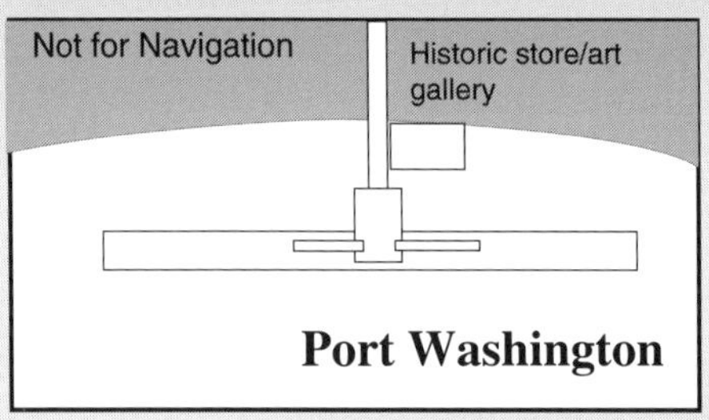

Port Washington,
North Pender Island
Transport Canada dock
Chart 3313, 3442
Manager •
Float length 45 m • Aircraft float • Public phone ashore • Nearby arts and crafts. Walking–island roads.

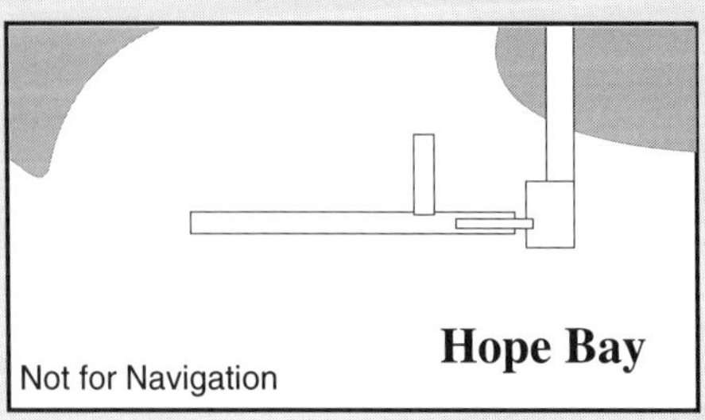

Hope Bay,
North Pender Island
Fisheries & Oceans dock
Phone (250) 629-4323
Chart 3313, 3477, 3461, 3462, 3441.
Manager • Float length 69 m • Lights • Public phone ashore • Adjacent shops.

Inset: Restaurant and pub overlooking Lyall Harbour at Saturna Landing.

Saturna Point

Charts 3313, 3473, 3462

Saturna Point Landing

& Fuel Dock. Lyall Harbour,
Saturna Island, BC. V0N 2Y0
Phone: (250) 539-5725
Harbour Master:
Gloria Manzano (250) 539-2229

Marina services:

Government dock. Limited space–mostly drop off/pick up. Not intended for overnight.
Fuel: Gas, diesel, outboard mix, oil.
Public phone ashore.
Store and restaurant ashore. Lighthouse Pub.

Customer services:

General store: fish tackle, hardware, licences, ice, bait. Propane available nearby.
Post office nearby–1 mile

Entertainment.

Annual lamb barbecue July 1st. at Winter Cove. Island tranquility. Eagles, herons, otters and seals.

Adjacent facilities:

BC Ferries. Use caution manouvering when ferry operating. Winter Cove anchorage, Boot Cove: Saturna Lodge and Restaurant. Vineyards on island–enquire at store.

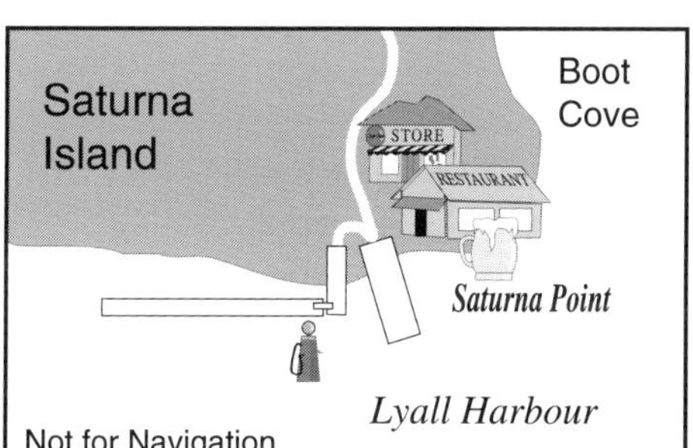

Lyall Harbour (Saturna Point)
Fisheries & Oceans dock
Chart 3310, 3477, 3462
Manager • Float length 62 m
• Lights •
Adjacent facilities include Fuel, restaurant/pub, store.

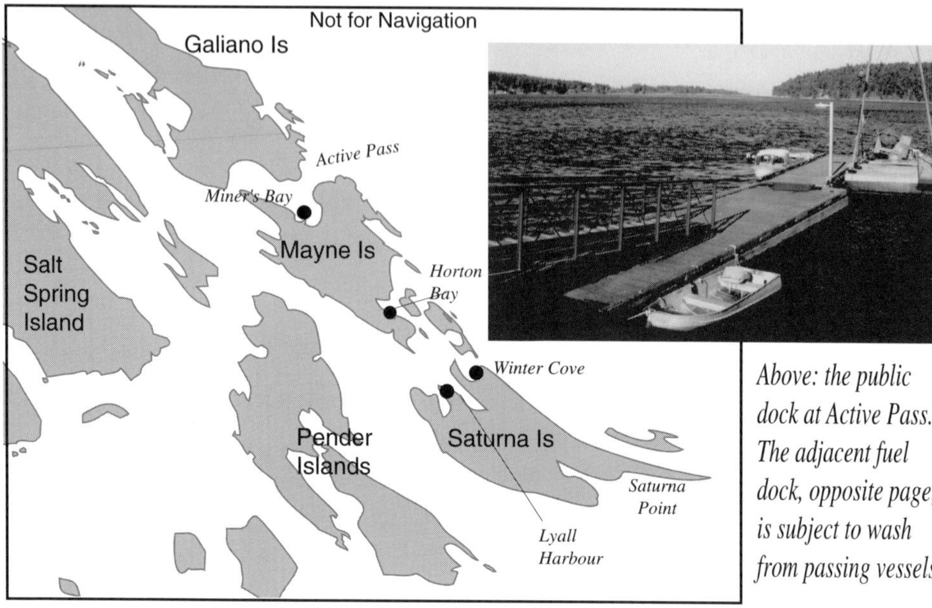

Above: the public dock at Active Pass. The adjacent fuel dock, opposite page, is subject to wash from passing vessels.

Active Pass

Scuba diving is popular in Active Pass. However one has to be mindful of the strong current and tide rips as well as the constantly passing ferries. It is best, therefore, if scuba divers use the services of professional dive charter operators rather than attempt diving the pass off their own private vessels.

Fishing in the pass, and particularly at each entrance to the pass is extremely popular and rewarding. However, here again, one should exercise caution due to the passage of ferries. Fishing vessels are obliged to move aside for approaching ferries. Common sense calls for such action to avoid collision and also to aid the ferries in their tight manouvering in the restrictive passage.

Left: BC ferry going through Active Pass. Opposite: The docks at Miner's Bay and the fuel float. It is open seven days a week for fuel. Other marine services are available through the general store on shore.

When approaching Miner's Bay for fuel, simply steer directly towards the fuel dock. The water shallows off towards the shore but mooring buoys indicate adequate water in their vicinity. Watch for the swells created by passing ferries and other vessels and wait for them to pass before attempting to dock at the fuel dock or the government floats behind the modest wood piling breakwater.

Charts 3313, 3473, 3462

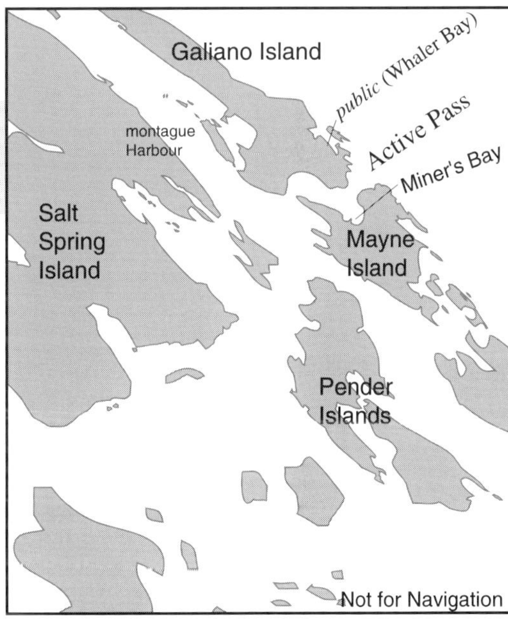

Miner's Bay

Mayne Island,
Miner's Bay

Transport Canada dock
Manager • Float length 37 m
Aircraft float • Lights •

Marine Services:

Fuel • gas, diesel, outboard mix. pop, confectionery. Tackle, bait, propane.
Adjacent restaurants, shops.
Moorage at wharf, slips.
No water or power on docks.

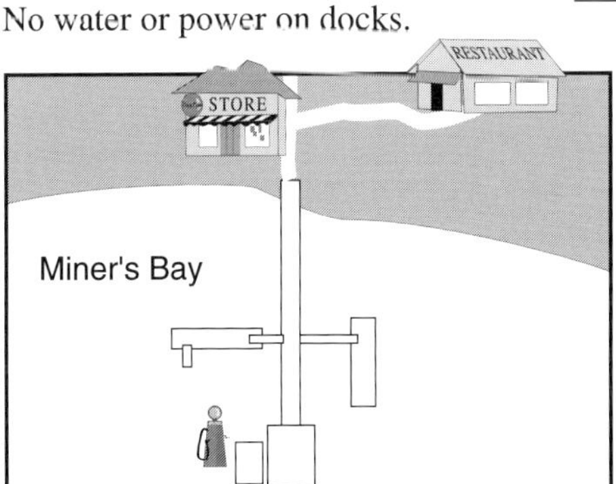

No laundry No washrooms.
Taxi service.
Access to settlement.
Near Ferry service–Gulf Islands to Mainland/Sidney.
Churches nearby.
Groceries, stores near.
Accommodation near.
Easy walking on island roadways.
Ferry wash causes some rolling at dock.

Galiano Island

Near the south end of Galiano Island Montague Harbour can be quite a busy anchorage in the summertime and although there are numerous mooring buoys as well as a fairly large dock, finding space to tie up can be difficult. Check in at Montague Harbour Marina where you can enjoy the ambience on the patio and a stroll to the nearby marine park. Or take in dinner at the Hummingbird Pub via a cheery local bus ride. Montague Harbour opens off Trincomali Channel and offers a safe haven for boats in all weather conditions. The odd violent storm has caused some discomfort at anchor in Montague but no known dangerous situations have been recorded. The dock at Montague Harbour Marine Park is operated and maintained by the Parks Department.

Moorage is charged for tying up overnight at that dock or on a mooring buoy. Access to the park is from the dock, which provides access to a delightful walk along the beaches or shoreline and the adjacent island. Views up Trincomali are spectacular in sunny weather and many people who have come by road and ferry to camp in the campgrounds also enjoy the scenery and the walking paths.

Fishing in Active Pass is popular but beware of the ferry traffic. It is illegal to impede the passage of a large vessel underway in designated traffic lanes, and plain foolish too. Watching wildlife in and around the pass, at Montague and across Trincomali in the inlets and bays of Prevost have been a favourite activity of mine and many of our boating friends for many years. Look for deer, eagles, cormorants and river otters.

Other places of interest, moorage and anchorages near Montague include Annette Inlet and Glenthorne Pass at Prevost Island, Long Harbour and Ganges on Salt Spring Island and Whaler Bay on the other side of Galiano from Montague. Prevost Island provides good anchoring in shallow water and sheltered coves. Annette Inlet is a large shallow inlet that can accommodate a large number of boats while Glenthorne Pass has a small bay at the end of a narrow passage which will handle several boats swinging at anchor. The passage itself was used at one time for anchoring to a greater extent than in recent times, however some permanently located crab traps and their attendant floats now deter boaters from mooring. The head of the inlet is the preferred anchorage.

The marine park dock at Montague Harbour. Boats anchor or use mooring buoys. The marina is located in the lower left in the aerial photograph at top.

Montague Harbour

Montague Harbour Marina

Bob and Rosemary Walker
RR #1 Galiano Island, BC. VHF 68
V0N 1P0 Charts 374, 3310
Phone: (250) 539-5733 Fax 539-2010
email: galiano@islandnet.com
Web: www.nwboat.com/montague

Marina services:

Fuel: Gas. Diesel. Marine supplies. Fishing licences, charts, bait, ice. Marine repairs available. Transient and permanent **moorage**. **Water**. **Power**: 30, 15 amp.

Customer services:

Store–Fresh produce, gift shop, island arts and crafts, books, charts, fishing licences, ice, bait and tackle. meals–light fare. Cafe. Draft beer. Liquor licence. No off-sales. Espresso bar–sundeck. Sea kayak float. Scuba diving good in Active Pass and nearby reefs. Use charter services. Public pay phone

Adjacent facilities and entertainment:

Marine park at Montague Harbour. Mooring buoys and dock. Extensive walks and camp ground, beaches and picnic facilities.

Public docks on Galiano Island:

North Galiano,
Galiano Island
(Spanish Hills)—Summer only
Fisheries & Oceans dock
Phone (250) 539-2352
Chart 3310, 3443, 3463
Manager • Float length 12 m
Public phone ashore •
Adjacent deli–Spanish Hills Store.

Retreat Cove, (photo opposite)
Galiano Island.
Fisheries & Oceans dock
Phone 539-2611
Chart 3310, 3442, 3463
Manager • Float length 24 m

Montague Harbour,
Galiano Island
Transport Canada dock
Chart 3310, 3473, 3462
Manager • Float length 50 m
Adjacent Montague Marina.

Above: The docks at Montague Harbour Marina.
Right: Store and expresso bar overlooking the docks.
Opposite: Note the expanded buildings in 1998 seen in the aerial photograph of Montague Harbour Marina. Watch for further additions and improvements.
Inset right: Retreat Cove.

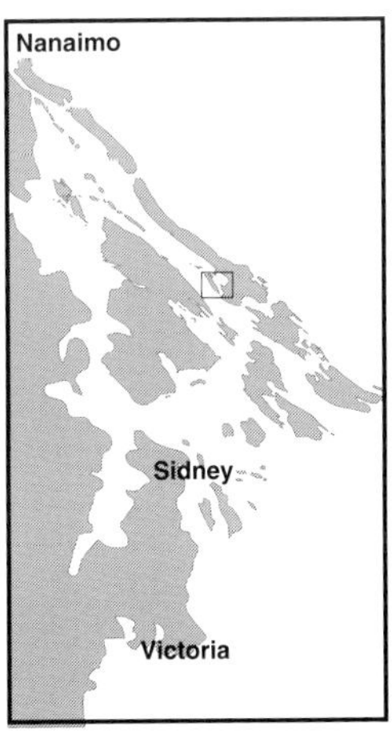

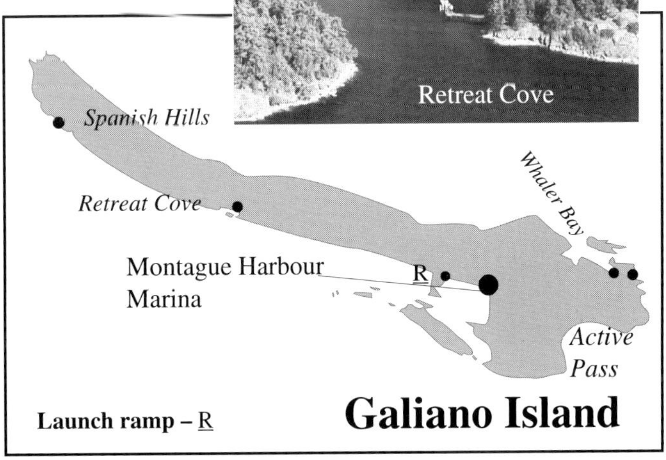

Pub and restaurant at Vesuvius on Salt Spring Island west side.

En route up the coast towards Thetis Island and its popular marinas many stop at Wallace Island. The dock (below) is in Conover Cove where you can go ashore and hike the easy trails that run the length of the island.

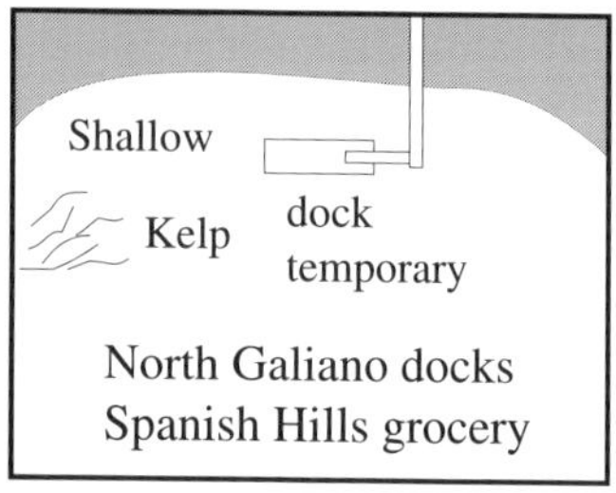

Bottom: At the dock in Horton Bay, Mayne Island.
Inset above: The store at the head of the dock at North Galiano (also called Spanish Hills).

Horton Bay,
Photo, left. Mayne Island.
Fisheries & Oceans dock
(250) 539-2402
Chart 3310, 3477, 3462
Manager • Float length 60 m
Garbage • Lights •

Telegraph Harbour occupies a Central location in the Gulf Islands. The famous "Cut" is located between the two islands.

Thetis Island

There's a waterway in the Gulf Islands that draws boats to the challenge of its shallows. One which lures sailors like the legendary sirens to an ignominious fate of running aground if not onto the rocks, to reach the prize beyond of sheltered anchorage and a fair haven from unexpected squalls and wind. The shallow, narrow passage that separates Thetis Island from Kuper is the eastern entrance to one of the most centrally located and popular anchorages in the Gulf Islands. The canal (known as 'The Cut') lets boats through only at medium to high tides and denies passage to all but the tiniest of craft at low tides. It dries at a one foot tide. But despite the quirks and whims of the famous passage, it is the waterway that experienced cruising yachtsmen associate with Telegraph Harbour. Explore the area: take a dinghy ride through the cut and see the shallows for yourself before taking your boat through. The alternative route into Telegraph is around the bottom of Kuper or the top of Thetis. If you are coming from a Vancouver Island base and returning to Vancouver Island after a stay at Telegraph Harbour, the passage is not an issue. But if you are crossing the Strait of Georgia and entering the Gulf Islands through Porlier Pass then the canal is the preferred access to the sheltered harbour. Choose a high tide to approach the canal or plan a longer, but pleasant detour around Thetis or Kuper.

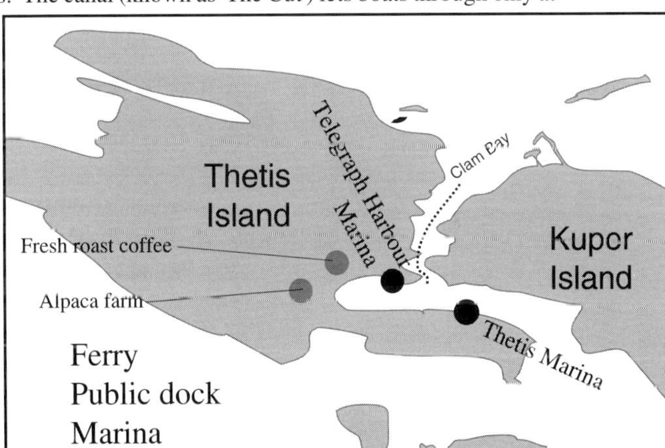

When you arrive in Telegraph Harbour, if you are among the average type of boat owner, the first thing you will do is find moorage, at Thetis Island Marina or at Telegraph Harbour Marina. At the former you may look for the pub, at the latter you will want to moor

Telegraph Harbour

Telegraph Harbour Marina

John, Jan & Julie Ohman
Thetis Island, BC,
V0R 2Y0
Phone: (250) 246-9511
Fax: (250) 246-2668

Marina services:

Fuel: Gas. Diesel. Oils. Outboard mix.
Repair service available.
Marine supplies. Fishing gear, licences, charts, bait, ice.
Marine repairs available.
3000 feet of moorage.
Water. Limited supply–use sparingly.
Power at docks: 30, 15 amp.
Multiple outlets.

Customer services:

Laundry, showers, washrooms for overnight moored guests.
Store/coffee shop, light meals, some groceries, milkshakes, ice cream, baked goods.

Charts 3477, 3313, 3442, 3463

Frozen foods, produce, gifts, arts and crafts, books, snacks. Continental breakfasts and light meals available in store or on patio.
Boating groups book events/rendezvous.
Playground. Picnic/barbecue facilities ashore. Book ahead in summer.

Nearby

Road access walking or cycling. Alpaca farm and Spinning Wheel. Donut bakery. Some nearby parkland and beach trails.

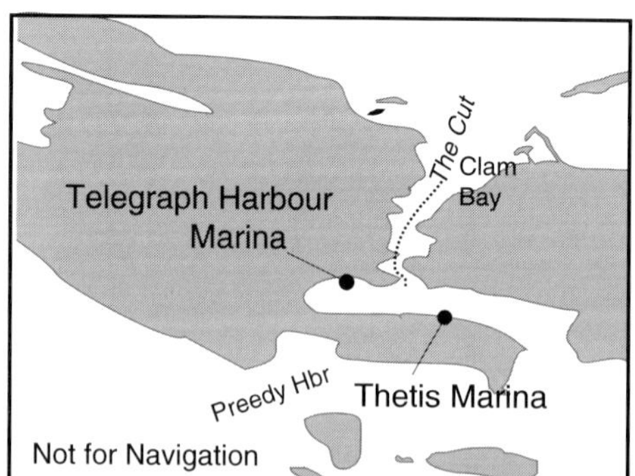

quickly and head up the dock for one of the delicious old fashioned milkshakes for which the marina store has become famous. Telegraph Harbour Marina is owned and operated by John and Jan Ohman, a most convivial couple who, like their neighbours Paul and Dawn Deacon at Thetis Island Marina, go to great lengths to welcome you and make you feel at home. In fact, so warm is their welcome that we have found ourselves cancelling our continuation plans in favour of just staying at the island longer than planned. And staying on is a logical choice considering the advantages of being there. The harbour is very protected from winds and weather and available moorage is usually plentiful even in the busy summer period.

Below: A friendly gathering place at Telegraph Harbour Marina.

When rounding Thetis Island the western entrance to Telegraph Harbour is via Preedy Harbour where seals can be seen sometimes sunning themselves on the rocks just off Foster Point. Thetis Island Marina juts out into the main passage and posted signs effectively call on boats entering Telegraph Harbour to slow down. Thetis Island Marina has a pub and serves meals from a more varied menu than that at Telegraph Harbour Marina. The regular clientele at the two marinas can be quite different, naturally, the pub being more of a social centre and congregating place for those who enjoy the pub atmosphere.

Telegraph Harbour Marina is suitable for families and family activities. Like most of the

---

Public pay phones ashore.

Daily float plane service. Short ferry trip to Chemainus shops and famous murals. Fresh roasted coffee available on island. Home/ boat delivery Asian cuisine, pizza. Enquire.

Entertainment.

Recreation building. Volleyball, shuffleboard, horseshoes.

Adjacent facilities:

Excellent bed & breakfast accommodations nearby. Alpaca hobby farm. Spinning. Knitted goods.

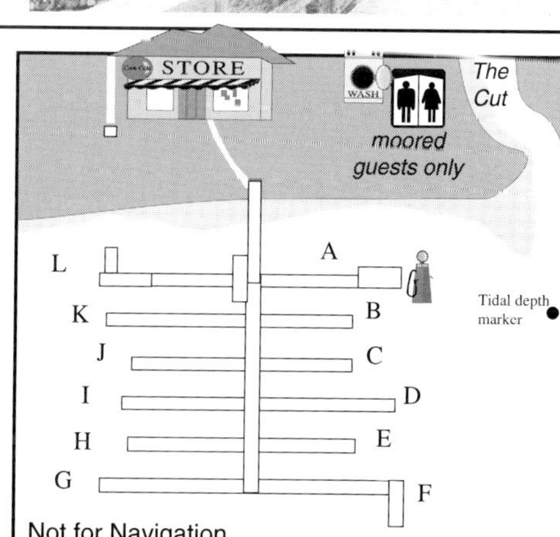

Thetis Island

Thetis Island Marina and Pub

Chart 3477, 3310, 3442, 3463
VHF 68

Paul and Dawn Deacon
Thetis Island Marina
General Delivery, Thetis Island
BC, V0R 2Y0
Phone: (250) 246-3464 fax 246-1433
email: *marina@seaside.net*
Web: www.seaside.net/homepage/marina/

Marina services:

Fuel: Gas, diesel, propane, kerosene, stove oil.

Marine supplies: Fishing licences, tackle, bait, charts, tide tables, ice. Marine repairs available.

Transient moorage.

Water. Limited supply–use sparingly.

Power at docks: 30, 15 amp. Multiple outlets.

Laundry, showers, washrooms.

Customer services:

Post Office. Store–groceries, dairy treats, ice cream. bakery products. Frozen foods, produce, gifts, arts and crafts, books, snacks.

Restaurant/pub–meals available inside or on large sunny patio.

Boating groups book weekend events/rendezvous. Playground. Picnic/barbecue facilities ashore.

Arrange/book ahead in summer.

Nearby church. Road access walking or cycling. Scuba diving arrangements and charters–ask marina for details.

Public pay phones ashore and at fuel dock.

Daily float plane service. Fresh roasted coffee available on island. Alpaca farm nearby.

Entertainment.

Short ferry trip to Chemainus shops and famous murals. Live entertainment in the pub on long weekends through summer months. Horseshoe pit, satellite TV. Darts. At the marina and nearby: Island arts and crafts, spinning, knitted goods.

Adjacent facilities:

Excellent bed & breakfast accommodations nearby.

Below: Thetis Island Marina protrudes into the passage between Thetis Island and Kuper Island. Opposite: Aerial view of Thetis Island Marina.

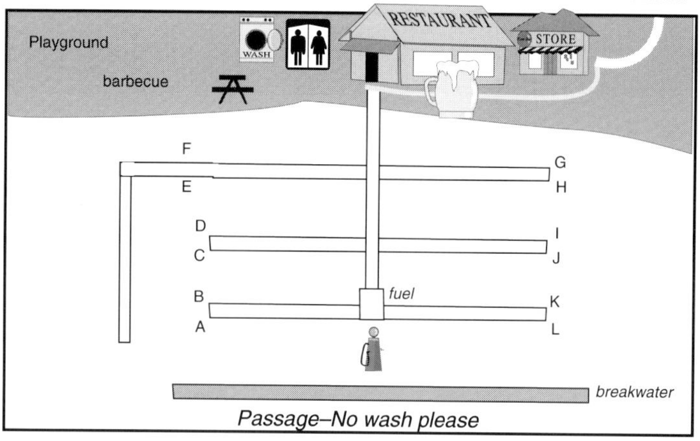

Gulf Islands water is in short supply on Thetis and boat owners are asked to use only what they need for refilling their fresh water tanks. Garbage is a problem for marinas but they do allow disposal of garbage by moorage customers. The Thetis Island post office and a propane filling station is located at Thetis Island Marina. The convenience stores at both marinas carry some souvenirs, charts and books as well as a selection of items for replenishment of boating supplies. Other than these stores at the two marinas there are no shops or shopping centres on the island. However, Chemainus on Vancouver Island, which is a short ferry ride away, has a selection of stores and restaurants to please everyone. It is worth the ferry ride to stroll around this artistic Vancouver Island centre. The passenger ferry leaves Thetis Island for the short run across Stuart Channel eight times a day.

Not far from Telegraph Harbour Marina is a well-known supplier of fresh roasted coffee—Pot of Gold is open at most times to sell their rich aromatic beans or freshly ground coffee to islanders and visitors alike. Just stroll up to the entrance of their property and make your purchases at the gate stall.

Thetis Island is known for its arts and crafts and not far from either marina is the Alpaca farm with its Spinning Wheel studio and crafts shop. Crafts on sale elsewhere on the island represent the works of various islanders and prices are generally more favourable than those for similar items in the cities. Look for their wares and details at the two marinas.

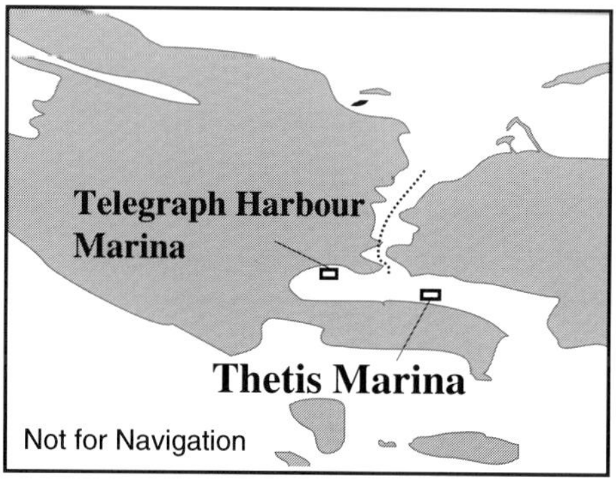

If you enjoy strolling a walk along any of the Thetis Island roads is relaxing and easy without any significant hills and traffic. Or at low tide you can don your boating boots and go beachcombing along the dry but marshy flats of the canal and watch your fellow boat owners trying their luck in the Cut.

Gabriola Island

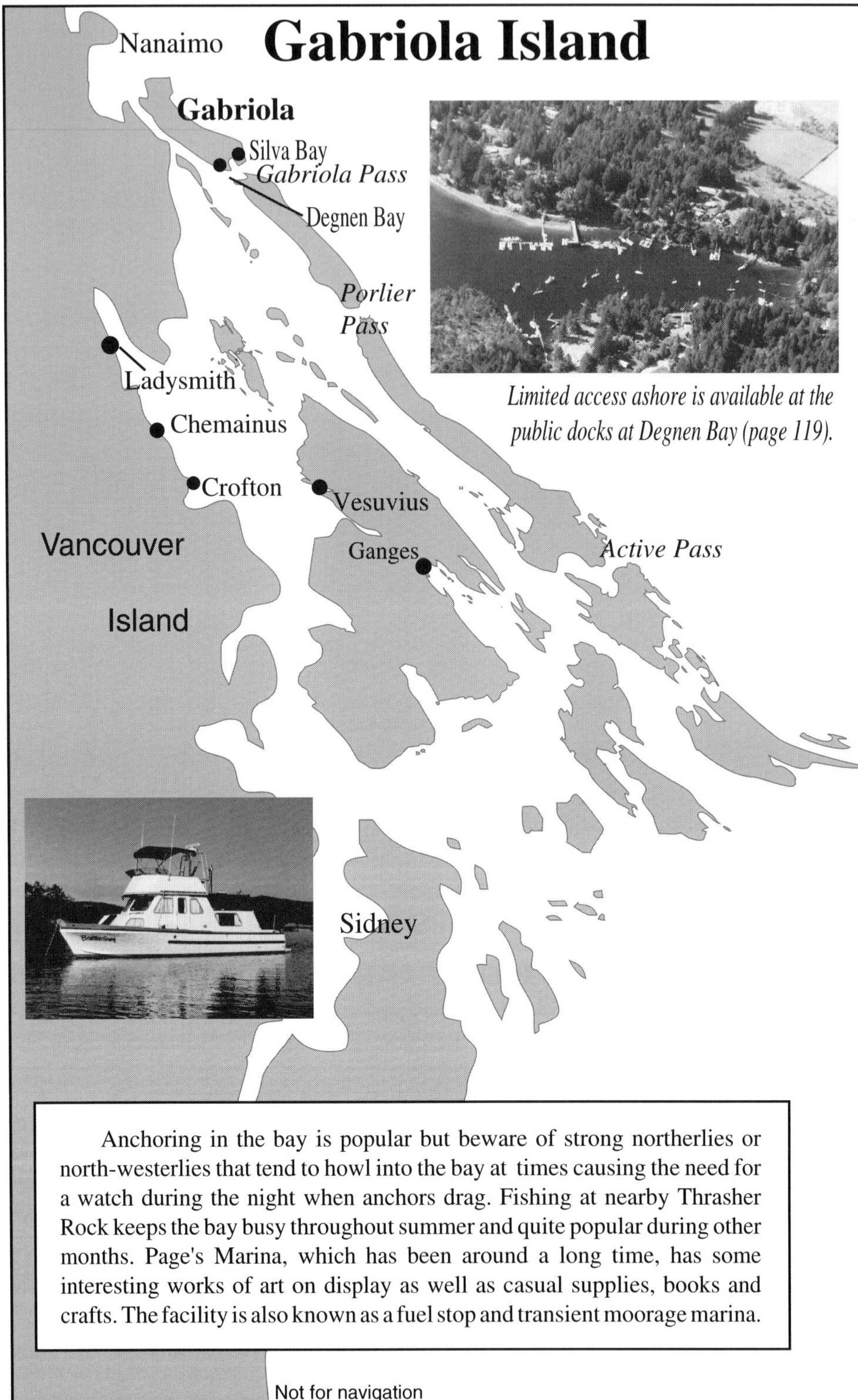

Limited access ashore is available at the public docks at Degnen Bay (page 119).

Anchoring in the bay is popular but beware of strong northerlies or north-westerlies that tend to howl into the bay at times causing the need for a watch during the night when anchors drag. Fishing at nearby Thrasher Rock keeps the bay busy throughout summer and quite popular during other months. Page's Marina, which has been around a long time, has some interesting works of art on display as well as casual supplies, books and crafts. The facility is also known as a fuel stop and transient moorage marina.

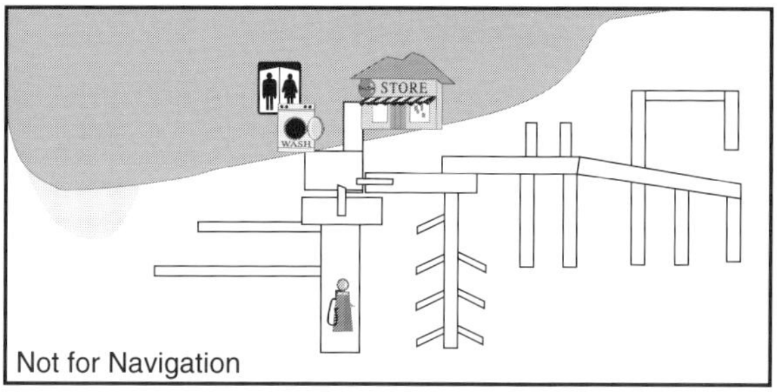

Silva Bay

1. Page's Marina

Ted and Phyllis Reeve
Site 30 RR #2, Gabriola, BC,
V0R 1X0
Phone: (250) 247-8931
Chart 3475, 3310.

Marina services:

Fuel: Gas. Diesel. outboard oil.
Moorage mostly to 30 feet, some larger.
Water at dock. Limited supply.
Power 15 amp.

Customer services:

Rental cottages. Picnic and campground. Office/store has art, books, etc.
Showers, laundry, washrooms.
Walking road access.
Good scuba diving in nearby locations. Dive shop ashore, rentals, air fills. (Nick Small).
Mobile repairs, service, available.
Public pay phone.

Entertainment.

Fishing charters can be arranged.

Adjacent facilities: Grocery store at Silva Bay Boatel. Fresh fish store.

Page's Marina dock has a large area and easy docking. Located at the end of the passage inside Sear Island. Watch depths at low tide and use a chart when navigating.
Page's has books and art at the office ashore.

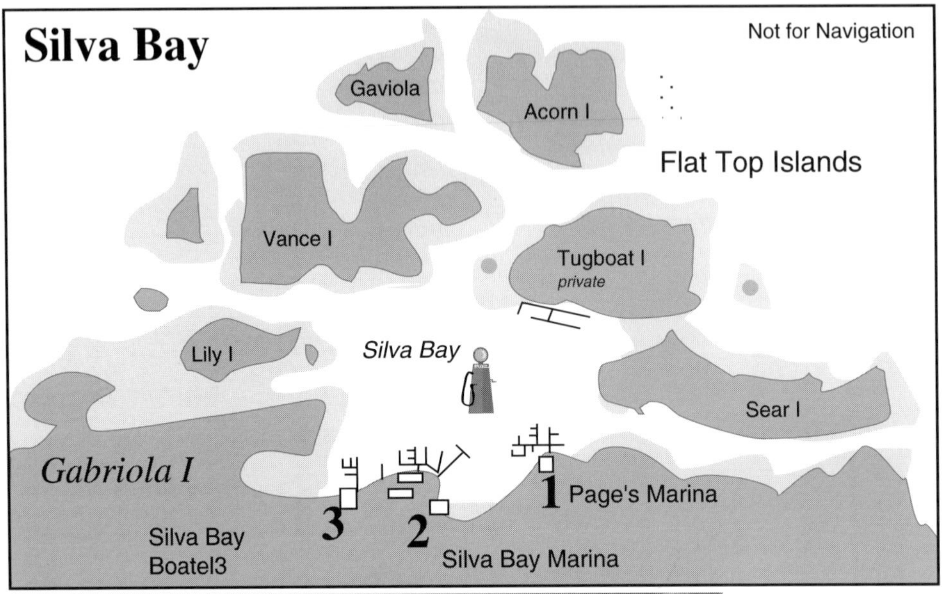

Aerial photograph shows location of Sterling Resorts Silva Bay Marina with the Silva Bay Boatel docks to its left.

2. Sterling Resorts Silva Bay Marina

Dock Manager
Gabriola, BC,
V0R 1X0
Phone: (250) 247-8662
Chart 3475, 3310, 3463, 3443.

Marina services:

Fuel: Gas. Diesel.Outboard mix.
Moorage, ways, shipyard.
Water at dock.
Power 15 amp. 30 amp.

Customer services:

Store, cafe, restaurant, pub–The Bitter End. **Washrooms, laundry, showers**, pool, spa, sauna.

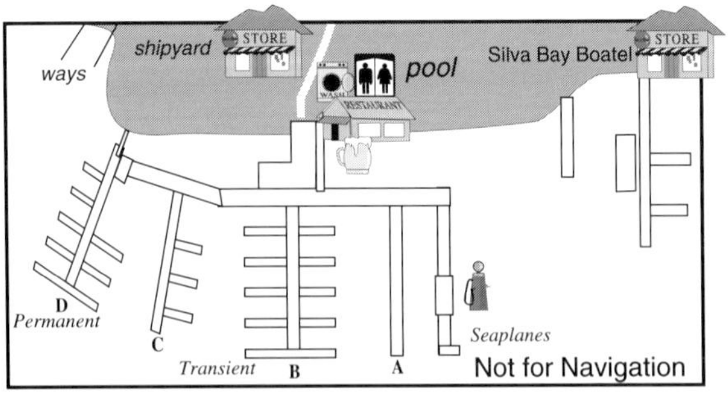

Silva Bay Boatel docks shallow at low tides. Limited space for tie-up.

3. Silva Bay Boatel

Chart 3475, 3310, 3463, 3443.

Audrey Leloupe
Silva Bay, Gabriola, BC,
V0R 1X0
Phone : (250) 247-9351

Marina services:

Moorage (limited–mostly small boats).

Customer services:

Grocery store. Fishing licences, bait, tackle, some hardware. Ice, propane. Head depot.

Walking road access.
Good scuba diving in nearby locations.
Public pay phone.

Entertainment.

Fishing charters can be arranged.

Adjacent facilities:

Accommodation in motel. Self-contained kitchenette suites–7 units.

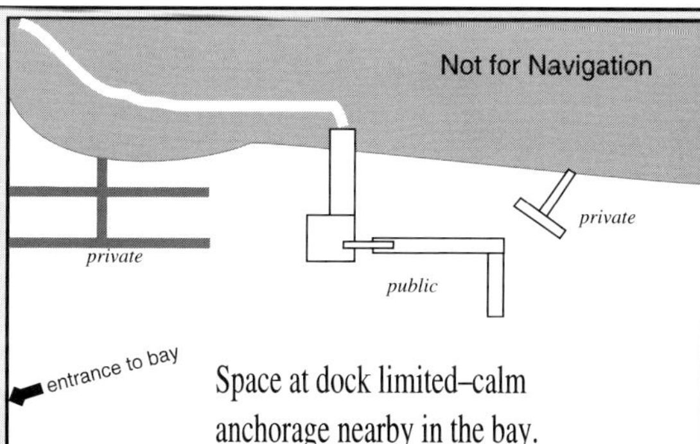

Space at dock limited–calm anchorage nearby in the bay.

Degnen Bay,

Gabriola Island public dock.
Fisheries & Oceans dock

Chart 3310, 3463, 3475
Float length 60 m
Garbage • Lights • Power
• Public phone •
Private docks adjacent.

Above: View of Newcastle Island from Nanaimo. Inset opposite page: The channel inside Newcastle Island is lined with marinas. Inset above: Newcastle Marine Park docks. Right: A ferry approaching the BC ferries dock in Departure Bay.

Nanaimo

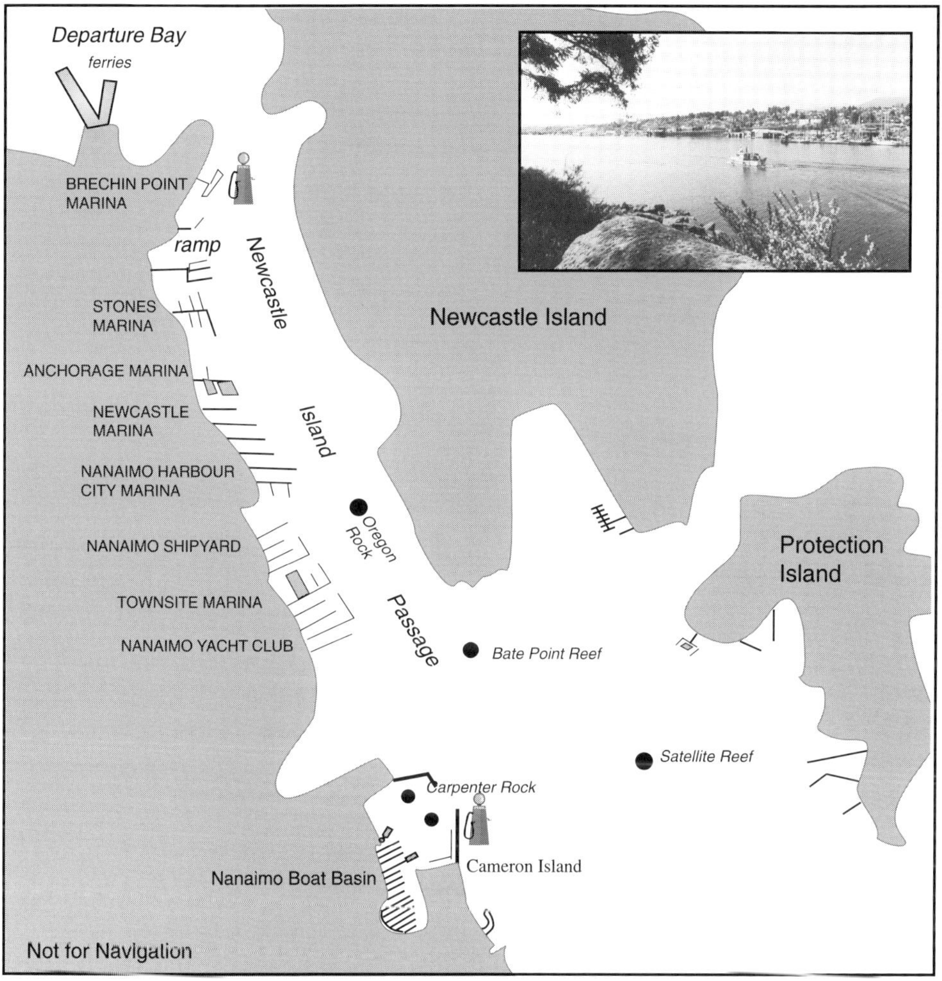

Nanaimo and its nearby islands.

Development of the city of Nanaimo has been fast and furious in recent years. Marinas and docks are plentiful with several having been upgraded substantially. The waterfront has undergone a massive face lift and new restaurants and public areas have evolved. There is a regular ferry service between Newcastle Island Marine Park and Nanaimo and another that serves the famous Dinghy Dock Pub on Protection Island. Their docks accommodate boats but space ashore is shared with the non-boating public.

Nanaimo

1. Nanaimo Boat Basin

Chart 3313, 3463, 3457/8 VHF 67

Dick Nickerson

Port of Nanaimo, BC,
Wharfinger (250) 754-5053
(and Cameron Island Marina)

Hazard: When proceeding past Oregon Rock use passage on Newcastle Island side, indicated by sign on mid-channel marker. Consult your chart.

Marina services:

Fuel, Petro Canada: Gas, diesel, mixed gas, ice. Service available.

Moorage. Large civic marina with pleasure boat moorage in summer. In winter docks are heavily used by fishermen. Reservations taken for new 600 foot floating breakwater pier for large vessels and adjacent **Cameron Island Marina** (phone **(250) 755-1216** June to August). Hydraulic crane to 500 kg.

Water at dock. **Power**: 15, 20, 30 amp. Also 50 and 100 amp.

Laundry, showers, washrooms.

Sani-station pump out.

Customer services:

Customs/phone 24 hour service.

Downtown Nanaimo at doorstep of marina. Road access walking or cycling or vehicle rentals. Public pay phones.

Daily float plane service.

Entertainment.

From restaurants, pubs and theatres to arts and crafts exhibits, stores and galleries.

Bathtub race every July. Many festivals and activities.

Adjacent facilities:

Walkway, plaza and shops. Shipyards, ways, all marine services. Lighthouse Bistro/Pub. Marinas. Newcastle Island Marine Park –docks, walking trails, camping, picnics, BC Ferries to mainland nearby. Walk on ferries to Newcastle or Protection Islands. Protection Island: Dinghy Dock pub. Anchor off and row to dock. Seats about 80 inside and on patio.

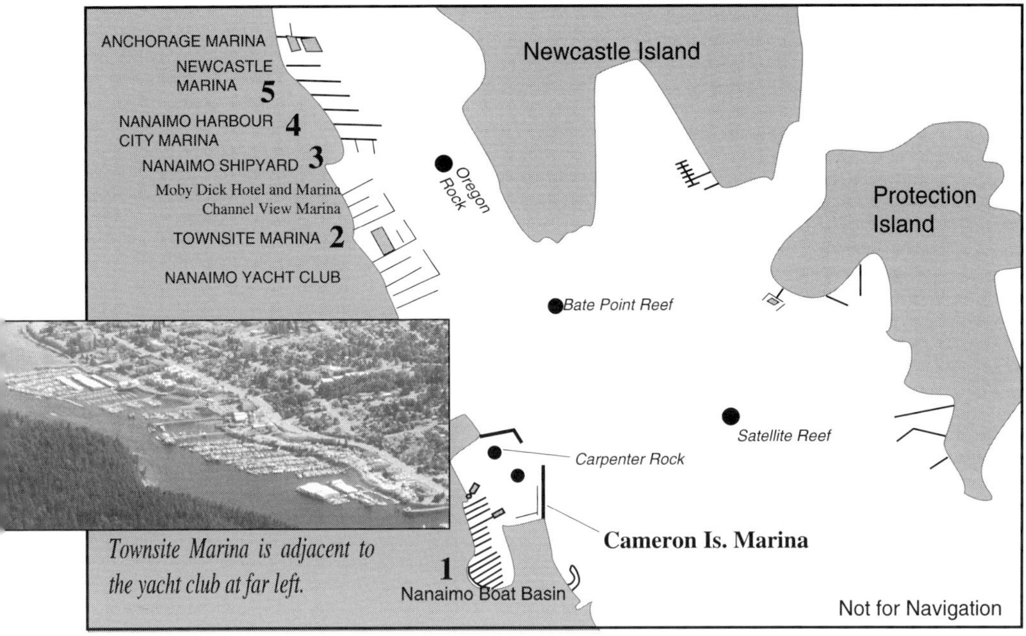

Townsite Marina is adjacent to the yacht club at far left.

2. Townsite Marina

Dock Manager

20 Townsite Road,

Nanaimo, BC,   V9R 4Z9

Phone (250) 716-8801 Fax (250) 716-7288

(see location on diagrams) **Call VHF   73**

Marina services:

Moorage boats 18' to 100 feet. Located adjacent Nanaimo Yacht Club.

Power 30, 50 amps. **Water.** Ice.

Washrooms, Showers, Laundry.

Power 30, 50 amps.

Nearby: Fuel docks, marine chandleries, Downtown Nanaimo. Restaurants, supermarkets, hotels, golf, museum, art gallery.

Note location of Townsite Marina alongside the yacht club.

Inset opposite: A fish boat cruises slowly down Newcastle Channel. No wake speeds are enforced. Mind the shallows in mid channel marked by a piling with a sign indicating correct passage.

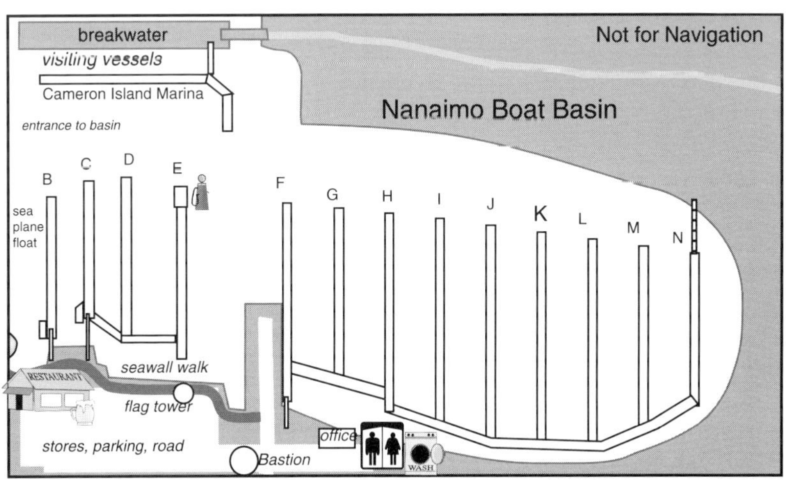

Dinghy Dock Floating Marine Pub

Hilda and Bob Banerd

Box 771, Nanaimo, #8 Pirate's Lane, Nanaimo, BC. V9R 5M2
Phone: (250) 753-2373
Charts 3457/8 3313, 3463. Call Dinghy Dock Pub on VHF 18A.

Marina Services

Moorage for overnight guests on a limited basis (as space allows). Tie up space for restaurant/pub.

Laundry, Showers. Fine dining. Walking on Protection Island. More moorage and anchoring at Newcastle Island.

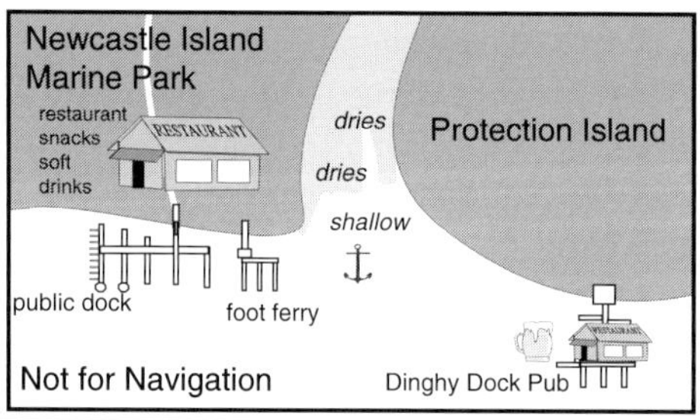

The Dinghy Dock Pub has won awards for excellence. It makes a perfect lunch or dinner stop and a limited number of boats will be able to stay overnight. The pub is built on floats and has slips to accommodate eight to ten small to medium sized boats. It is best to anchor out and come to the pub by dinghy. Or, if you are spending the night at Nanaimo take the ferry that leaves regularly from the Boat Basin. Facilities at the Dinghy Dock Pub include showers, laundry and washrooms.

Newcastle Island Marine Park is a short distance away and there are several slips for medium to larger sized boats as well as numerous slips for smaller boats and dinghies. Going ashore at Newcastle Island is a treat, providing lots of treed pathways for hiking, walking or cycling. An interpretive centre functions in summer for the use of students, groups and others interested in the use of the facilities.

The ferry between the Nanaimo Boat Basin and Protection Island leaves the docks hourly, 9:00 am to 11:00 pm.

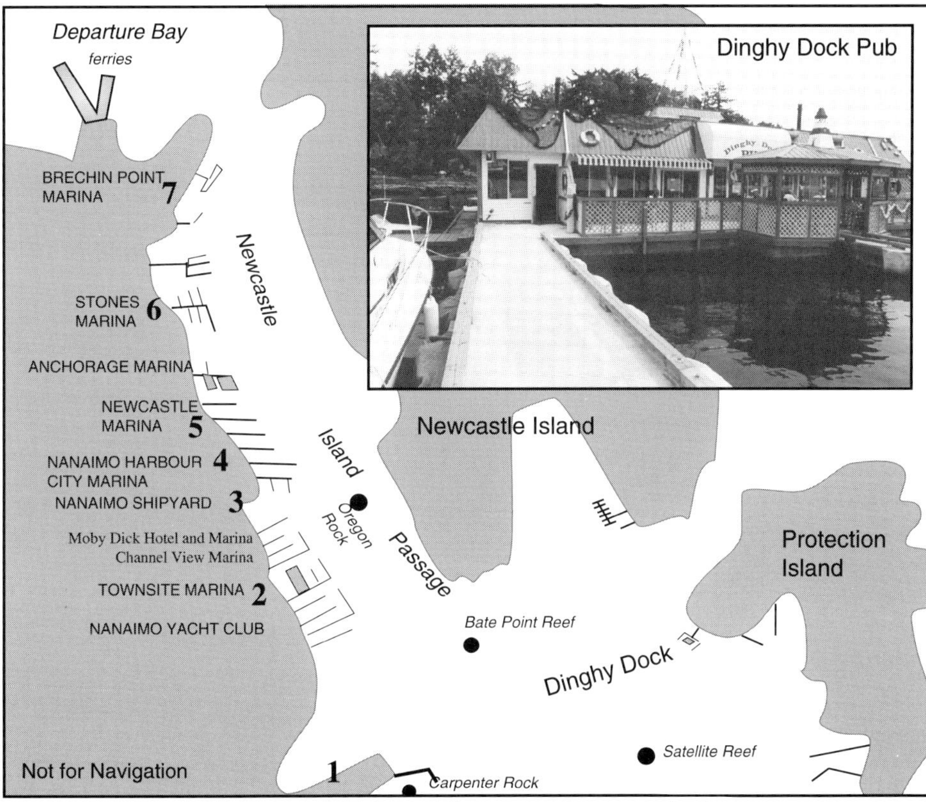

Dinghy Dock Pub (inset). Opposite: Aerial view of location on Protection Island.

3. Nanaimo Shipyard

1040 Stewart Avenue
Nanaimo, BC., V9S 4C9
Phone: (250) 753-1151
Fax: (250) 753-2235
Chart 3313, 3463, 3457/8
email: *rvw@island.net*
Web: *www.island.net/~nsy*

Hazard: Enter Nanaimo via south of Protection Island. From the north/Departure Bay: use correct channel when passing the mid channel rock.

Passage on Newcastle Island side.

Marina services:

Moorage. Haulouts. Repairs. Vessels to 200 feet.

Chandlery Phone: (250) 753-1244.

All boating supplies.

Water at dock. **Power** at docks.

Customer services:

Downtown Nanaimo near marina. All facilities and services are available.

Road access walking or cycling.

Vehicle rentals.

Daily float plane service nearby.

Entertainment.

From restaurants, pubs and theatres to arts and crafts exhibits, stores and galleries at Nanaimo. Bathtub race every July.

Adjacent facilities:

Shipyards, all marine services.

Nanaimo Yacht Club reciprocal moorage.

Marinas.

Newcastle Island Marine Park; docks, walking trails, camping, picnics.

BC Ferries to Vancouver.

Ferry to Galiano Island, Newcastle Island and Protection Island.

4. Nanaimo Harbour

City Marina

Ron and Susan Mielke
1250 Stewart Avenue,
Nanaimo, BC., V9S 4C9
Phone: (250) 754-2732 Fax: 754-7140
Chart 3313, 3463, 3457/8

Hazard: Enter Nanaimo via south of Protection Island.

From the north/Departure Bay watch correct channel when passing the mid channel rock. Passage on Newcastle Island side.

Marina services:

Moorage. Mostly permanent. Haulouts. Repairs, welding. Boatyard. Travel lift.
Water at dock. Ice.
Power at docks: 15, 20 amp
Showers, washrooms.

Customer services:

Downtown Nanaimo near marina. All facilities and services are available.
Road access walking or cycling or vehicle rentals.
Public pay phones.
Daily float plane service.

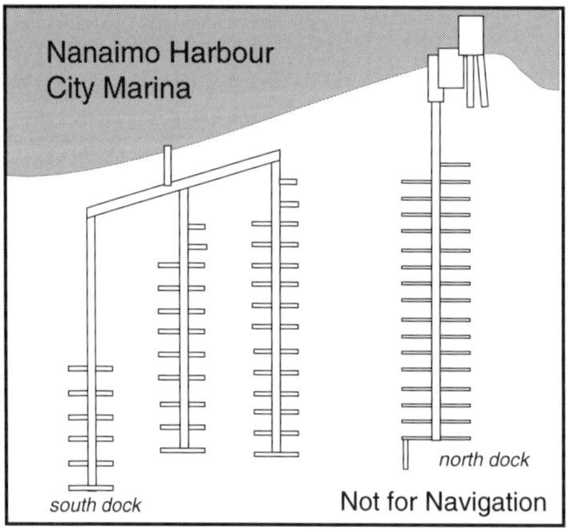

Entertainment.

From restaurants, pubs and theatres to arts and crafts exhibits, stores and galleries. Bathtub race every July.

Adjacent facilities:

Shipyards, all marine services.
Newcastle Island Marine Park; docks, walking trails, camping, picnics.
BC Ferries to Vancouver. and Galiano.
Ferry to Newcastle and Protection Island.

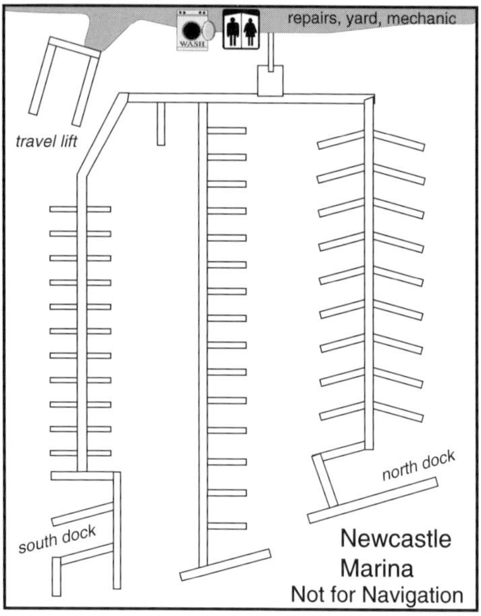

5. Newcastle Marina

1300 Stewart Avenue
Nanaimo, BC., V9S 4E1
Phone: (250) 753-1431
Fax: (250) 753-2974
Chart 3313, 3463, 3457/8
email: *nmarina@bookmark.com*

Hazard: Enter Nanaimo via south of Protection Island. From the north/Departure Bay: use correct channel when passing the mid channel rock.

Passage on Newcastle Island side.

Marina services:

Moorage. Haulouts. Repairs. Welding. Yard. Storage. Mostly permanent moorage.
Water at dock. **Power** at docks.

Customer services:

Downtown Nanaimo near marina.
All facilities and services are available.
Road access walking or cycling.

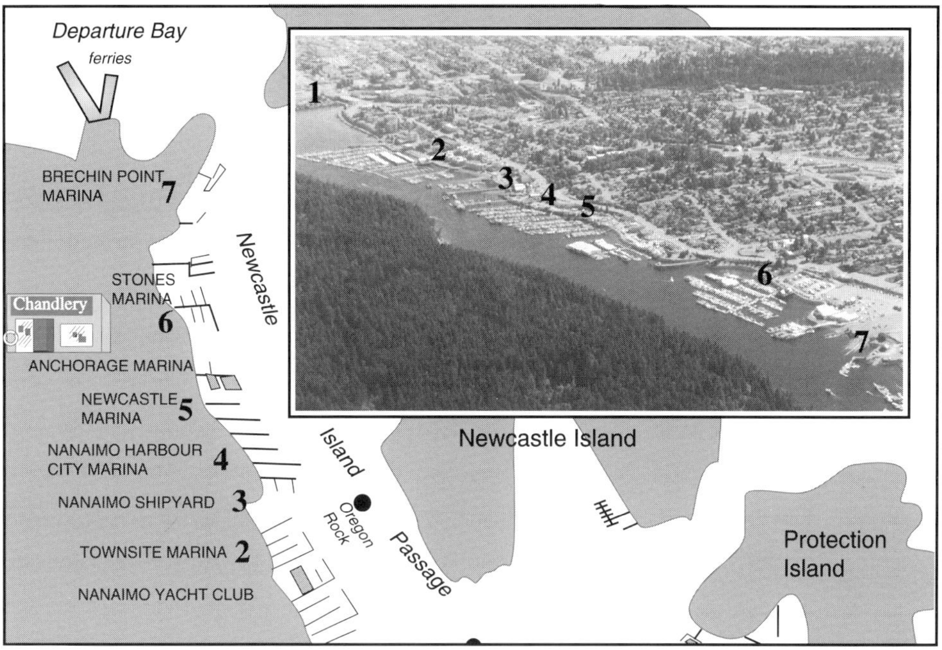

6. Stones Marina

& RV Park

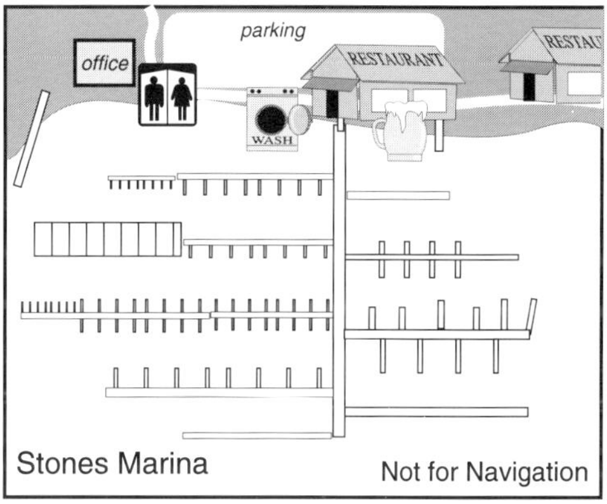

Mark Stone
1690 Stewart Ave., Nanaimo, BC.
V0N 2H0
Phone: (250) 753-4232
Chart 3313, 3463, 3457/8
email: *stones@island.net*
Location: next to Seaquarium, and near BC Ferries. **Launch ramp adjacent.**

Marina services:

Moorage. Transient in summer months.
Power: 15, 30 amp
Water at dock.

Customer services:

Ice. Laundry, showers, washrooms
Pay phone. Garbage disposal. Store. Tackle, fishing licences, bait. Mobile LPG service. Marine sales. **Pub and restaurant.**
Walking: Road access walking or cycling.

Entertainment.

Nearby access to ferries to Galiano Island, Newcastle and Protection Islands.

Adjacent facilities:

Muddy Waters Pub, restaurant and RV Park. Brechin Point Marina fuel dock and seaplane float–nearby.

7. Brechin Point Marina

Ph: (250) 753-6122.
Emergency only: After hours Chris at (250) 758-3646 or Doug at (250) 754-2317.
email: *landahl@island.net*
Customs, Seaplanes, marina.
Fuel: Diesel, gas, oils, propane.

Crossing the Strait

Silva Bay is probably your departure point of choice if you are journeying from the Gulf Islands to Howe Sound or to the Sunshine Coast, simply because of its relative close proximity to Vancouver. The closest landfall on the mainland is the entrance to the north arm of the Fraser River at Point Grey. If you travel up the river you will find moorage and fuel at Richmond. Some swift currents in the river when the tide is falling can make manoeuvering at docks tricky so exercise caution. Tugs travelling up and down the channel cause large washes. Be cautious when crossing over them.

Farther up the river is the confluence of the north arm with the main arm of the Fraser and New Westminster. There is a possible stop at New Westminster at the public market but the currents mostly keep boats from an easy docking and even when you have docked there is no easy access ashore. Up the Fraser to Pitt River you pass under a very low bridge to Pitt Lake. If you have a smaller, fast boat and decide to do this trip check chart #3062 and ensure you can pass beneath the bridge. Mind the shallows and stick to the chart. Fuel is available at the confluence of the Pitt River and the Alhouette and the marina may be able to accommodate a small boat overnight. There are pleasant temporary anchorages in Pitt Lake and one marina located on the south western shore. Winds come up suddenly and the lake can be quite uncomfortable if you are anchored in the wrong place.

Top: The lightship at Sand Heads, unmanned it provides no wave height information.
Above: The coast guard hovercraft running along the Fraser River.

Had you entered the Fraser via the main arm you would have passed the Sandheads Lightship. Hopefully you monitored the weather and checked the tides, for a passage of the river mouth at the wrong time can be dangerous.

From Silva Bay to Vancouver Harbour or Howe Sound is one of the shortest crossing of the Strait other than between the lightship at Sandheads and Porlier Pass or between Point Roberts and Active Pass/East Point. Howe Sound is known for its excellent moorage in close proximity to Vancouver.

Downtown Vancouver nowadays has plenty of transient moorage. The main visitors' dock at Coal Harbour Marina.

Centre: Crescent Marina is tucked in the south curve of Boundary Bay, behind the railway line.

Left: Looking east across False Creek from Granville Island's Pelican Bay.

Vancouver Area

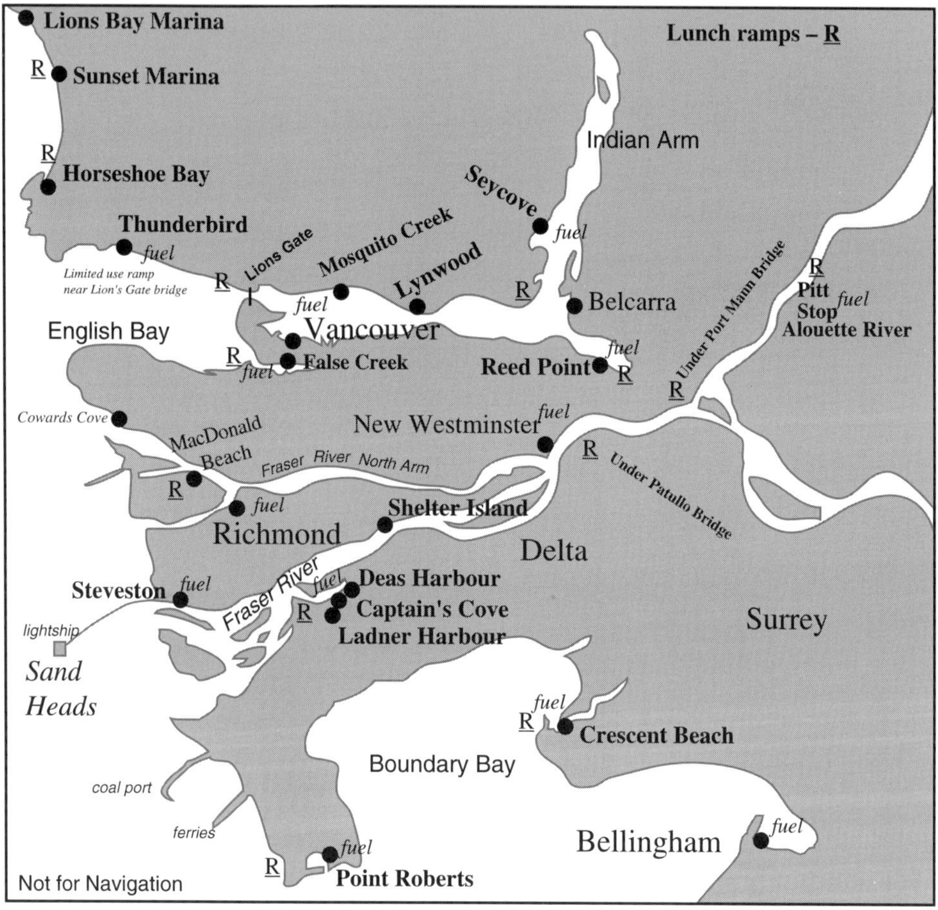

Marinas in the Greater Vancouver area do offer moorage for overnight customers. However, most of them are dedicated to providing permanent moorage and although some have many slips they usually can allocate space as needed during summer when many of their customers' boats are away. Because of the river estuary and large delta making up most of the area, marinas are located at some off the beaten track locations, such as up the Fraser River. Boat operators in transit from southern ports to places such as the Sunshine Coast and Desolation Sound rarely want to take the time to deviate from their course to an up-river marina just for an overnight stop. So there should be some purpose in cruising up the river, either the main arm of the Fraser or the north arm. A fast powerboat will have no trouble (other than debris and logs) accessing some up-river marinas, but slower boats looking simply for an overnight mooring should consider Vancouver's downtown area where there are many attractions. In the Fraser River, one can cruise all thc way up to the Albion Ferry Crossing safely by just staying in mid channel. It is possible to reach Mission or farther up river with the help of a river pilot. Or turn off on the Pitt River and enjoy a splendid run into Pitt Lake. There is fuel on the

1. Steveston

Near the mouth of the Fraser River. Fisheries & Oceans dock Gulf of Georgia *(photograph above)*
Ph/fax (604) 277-4712 Chevron dock.
Charts 3490, 3601, 3463
Manager • Float length 778 m
Breakwater • Grid • Garbage • Waste oil disposal • Parking •
Water • Lights • Power • Public phone ashore • Washrooms • Showers •
Adjacent city restaurants, shops, marine stores, chandleries, repair facilities. Auxiliary Coast Guard. **Fuel** at three locations. Additional public docks (Paramount) for fishing and commercial vessels. **Launch ramp.**

Crescent Beach

Fisheries & Oceans dock
Beach float
Chart 3463
Manager • Float length 12 m
Lights •

2. Captain's Cove Marina

Gina Tweed
6100 Ferry Road, Ladner, BC. V4K 3M9
Phone: (604) 946-1244

Chart 3463

Fresh water moorage at visitor's dock. **Water. Power** 15, 30 amps. **Fuel:** gas, diesel, engine and outboard oils. **Showers, laundry, washrooms.** Public phones. Rusty Anchor Pub adjacent. Travel lift 30 ton, repairs, service, storage and workyard, power wash, painting. **Launch ramp** adjacent.

3. River House Marina

5825 60th Ave, Ladner, BC. V4K 3E2
Phone (604) 940-4496

Chart 3490

Limited moorage (mostly permanent).
Some water, power available–15 amps.
Public phones.
River House restaurant and marine pub.
Low bridge beyond Captain's Cove Marina 14' clearance at medium high tide.

Pitt River at **Pitt Meadows Marina** (and launch ramp) at the confluence of Pitt River and the Alouette River.

Entering the Fraser River is safe provided you do so when winds are relatively calm and the tide is right. At about maximum low tide with a northwesterly wind you are well advised to wait. If you are crossing from Active Pass or Porlier, or coming up from the States it is best to wait before approaching Sand Heads until the wind drops and the tide slackens. Nearing high tide in calm to light winds is the best time to enter the river. Beware of the shallows off the mouth of the river, especially if you do find yourself running into some wind and waves. It is well marked but the shallows actually extend somewhat beyond the markers, especially around the beacon to the south of the lightship.

Stopping at Steveston (public docks) is a treat if you find moorage and go ashore. There the variety of shops, cafes and restaurants will keep you busy for hours. However,

4. Crescent Beach Marina (1967) Ltd.

B.J. Chapman
12555 Crescent Road, Surrey,
BC. V4A 2V4
Phone (604) 538-9666 Fax 538-6367
Chart 3463. See map on page 126.
Entrance: Follow markers in long channel through Boundary Bay, using red right returning. Moorage (phone for reservations). Water, power 15 amp. **Washrooms,** public phones. **Fuel:** gas, diesel. Ice. Chandlery, bait, fishing supplies. Haulout up to 45' power, 30' sail. **Launch ramp.**

5. Shelter Island Marina

Glenn Rae
120–6911 Graybar, Richmond,
BC. V6W 1H3 Ph: 1-877-270-6272
Phone: (604) 270-6272 Fax: 273-6282
Charts 3463 3490
Fresh water moorage (permanent). **Water. Power. Showers, laundry, washrooms.** Public phones. Marine hardware, repairs, service. Two travel lifts to 150 tons–vessels to over 130', 30' beam. Workyard for up to 500 boats. Beer and wine store. Restaurant and pub adjacent.

6. Bridgepoint Marina

Gary Cross
8831 River Road, Richmond,
BC. V6X 1Y6
Phone (604) 273-8560 Fax: 231-8081
Chart 3490
Fresh water moorage. Transient welcome. **Water, power, showers, laundry, washrooms**. Public phones. Pub, restaurant on site. Sales adjacent. Fuel dock nearby.

7. Delta Vancouver Airport Hotel Marina

Bob Dunne
3500 Cessna Drive, Richmond,
BC. V7B 1C7 Chart 3490
Phone: (604) 278-1241 Hotel & Marina
Phone: (604) 273-4211 Delta Charters
Fax Marina: (604) 276-1975
Transient moorage. Water, power.
Public phones. Hotel facilities and services. Restaurant. Near Vancouver International Airport, Richmond city facilities, restaurants and shopping centres.

---

there is extremely limited moorage for pleasure craft when the fishing fleet is in. One dock has been allocated for pleasure boat tie up, and that is at the outside dock of the marina after you pass the two first fuel docks. Other moorage is available by rafting up to a fish boat on the inside of that dock. The harbourmaster can be called at (604) 271-6807 and if there is alternative moorage he will advise you. This is a major commercial port and a good place to fuel up.

Going farther up the river there is Captain's Cove Marina for haulouts, repairs and service as well as some overnight moorage, fuel, ice and a pub, The Rusty Anchor, which serves good meals seven days a week. Beyond Captain's Cove is Deas Harbour with its River House restaurant and marina beyond the low bridge of Highway 99. The restaurant boasts a splendid pub and patio service. The village of Ladner is tucked away down Ladner Reach which opens off the Fraser at the entrance to Captain's Cove. Navigate with caution and move slowly past waterfront homes. A large public dock at Ladner welcomes mariners, especially when the fishing fleet is away. Phone (604) 946-8430 for information. Another good facility with marine pub and restaurant is at Shelter Island a little farther up river. Here the currents can be quite strong and tie-up limited. But if you

8. Vancouver Marina

John Short
8331 River Road, Richmond,
BC. V6X 1Y1
Phone: (604) 278-9787 Fax (604) 278-1738
Chart 3490
No transient moorage. Permanent up to 45' open, 38' covered, Water, power. **Fuel:** gas, diesel, stove oil, engine and outboard oils. Public phones. Repairs, service. Restaurant. Near Richmond city facilities, restaurants.

9. Quayside Marina

900 Quayside Drive,
New Westminster. V3M 6G1
Phone (604) 520-1776 Fax (604) 520-5645
Chart 3490. Go beyond on **3489, 3488.**
Limited visitor moorage. **Water. Power**. Hotel, restaurant, public market.

10. Skyline Marina

Dave Shirley
8031 River Road, Richmond,
BC. V6X 1X8
Phone: (604) 273-3977
Chart 3490
Shipwrights. Some transient moorage, water, power–15 amp. Fuel (nearby): gas, diesel, stove oil, engine and outboard oils. Public phones. Repairs, service, chandlery nearby. Nearby Richmond city facilities, restaurants.

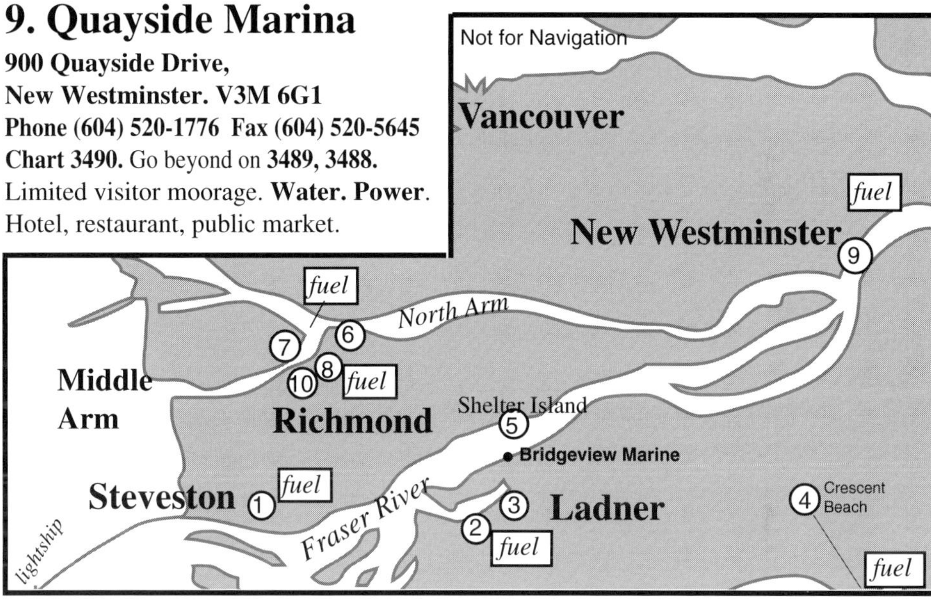

need a haulout for a large boat or a workyard in which to block it up Shelter Island Marina can provide the necessary facilities.

You could cruise on to New Westminster and return to the Strait of Georgia by going down the north arm of the Fraser. Part way down you will come to Richmond's marinas just after passing under Knight Street bridge and Oak Street bridge. Fuel is available at Vancouver Marina just beyond the swivel bridge on the Middle Arm. There is another fuel dock as you continue down the river, directly beneath the newer Arthur Laing bridge.

A stop worth considering for a lunch hour or pleasant day in the sun is MacDonald slough which opens to the south as you pass down-river of MacDonald Beach. It is a shallow slough but with a fairly constant and adequate depth. There are log booms to tie up alongside and one can observe a variety of water fowl that inhabit the area. At the mouth of the North Arm is a tiny cove called Boat Basin or Coward's Cove. It is where mariners sometimes sit and wait for a change in conditions before venturing out into the Strait. Here again, log booms serve for tying up, or it is safe to anchor. The breakwater and nearby Wreck Beach are popular among nature lovers.

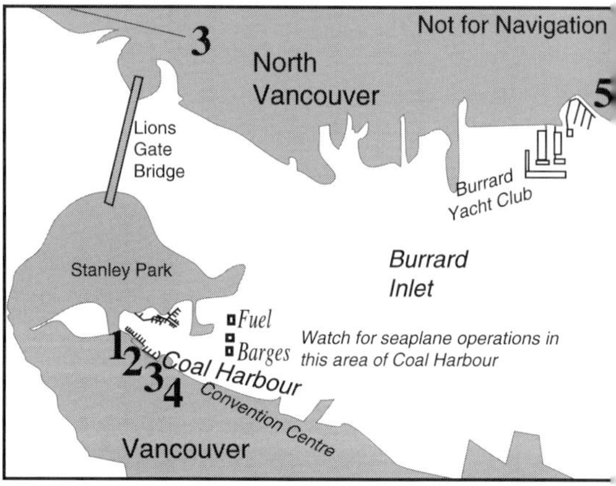

1. Harbour Ferries

Peggy Walker
#1 North foot Denman,
Vancouver,
BC., V6G 2W9

Phone: (604) 687-9558 Fax: (604) 687-5868

Moorage. **Water. Power** 15, 20, 30 amp. Fuel (nearby): gas, diesel, stove oil, engine and outboard oils. **Washrooms.** Public phones. Chandlery, repairs, service, nearby. Located next to Stanley Park and charter vessel basin.

2. Westin Bayshore Marina

Jane Boddy
1601 West Georgia, Vancouver,
BC. V6G 2V4
Phone: (604) 682-3377 Fax: (604) 687-3102

Moorage to approx 100 feet, **Water. Power** 30, 50 amp. Fuel (nearby): gas, diesel, stove oil, engine and outboard oils. Hotel facilities incl: showers, laundry, washrooms, health club, spa, pool. Public phones. Chandlery, repairs, service, Stanley Park nearby.
Near city centre, restaurants, facilities.

3. Thunderbird Marina(s)

5776 Marine Drive, West Vancouver,
BC, V7W 2S2
Phone: (604) 921-7434 Fax: (604) 921-7486
Charts 3311, 3534, 3481

Limited moorage, **water, power**.
Fuel dock (nearby): gas, diesel, stove oil, engine and outboard oils. Public phones. Marine store, tackle, bait, fishing licences, repairs, service. 25 ton travel lift for boats to 51 feet. Workyard, storage.

Also downtown Vancouver moorage:
Bayshore (West) Marina. Guest moorage, **water, power,** phones. Near hotel and downtown amenities. Near Stanley Park.

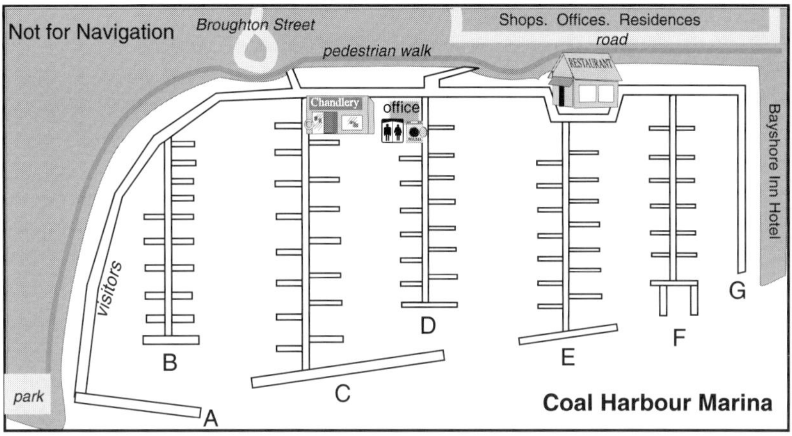

Photo, opposite, shows the large basin containing Coal Harbour Marina. To its right are Westin Bayshore, Harbour Ferries and Thunderbird Marina.

Use charts 3311, 3495

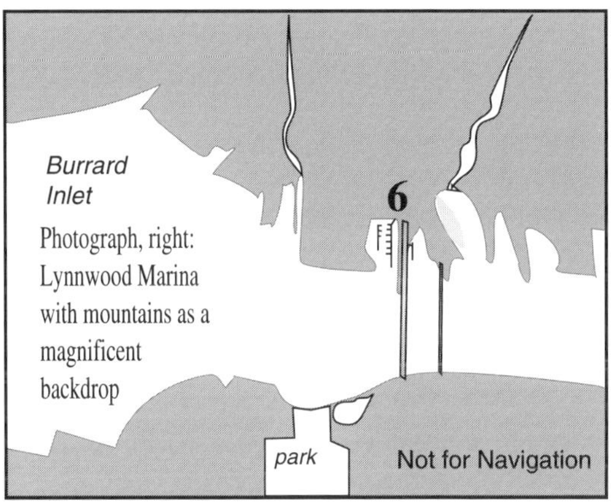

4. Coal Harbour Marina

John Nicholls
1525 Coal Harbour Quay,
Vancouver, BC, V6G 3E7
Phone: (604) 681-2628 Fax: 681-4666
email: *coalhbr@seatosky.com*
Web: *www.seatosky.com/coalhbr*

Marina Services VHF 68

Moorage. Cable vision. **Washrooms. Water. Power** 30, 50, 100, 150 amp.
Fuel (nearby): gas, diesel, stove oil, engine and outboard oils. Vacuum sewage system. Public phones. Wright Mariner Supply–chandlery. Repairs, service–nearby.
Coffee shop, convenience store.
Adjacent to city centre, restaurants, shops.

5. Mosquito Creek

Bill Williams
Ft Forbes Ave., North Vancouver,
BC. V7L 4J5
Phone: (604) 987-4113 Fax: (604) 987-6852
Moorage, water, power. F**uel:** gas, diesel. Washrooms. Public phones. Marine supplies, repairs, service. Crane. 35 ton travel lift. Near North Vancouver restaurants, shops.

6. Lynnwood Marina

Roger Gibson
1681 Columbia, North Vancouver,
BC. V7J 1A5
Phone: (604) 985-1533 Fax: (604) 985-8892

Moorage, **water, power**. Ice, bait, fishing tackle, licences. **Washrooms**. Public phones. Chandlery, repairs, service. Painting, storage and work yard, 60 ton travel lift to boats 70' by 18' size.
Restaurant. Services.

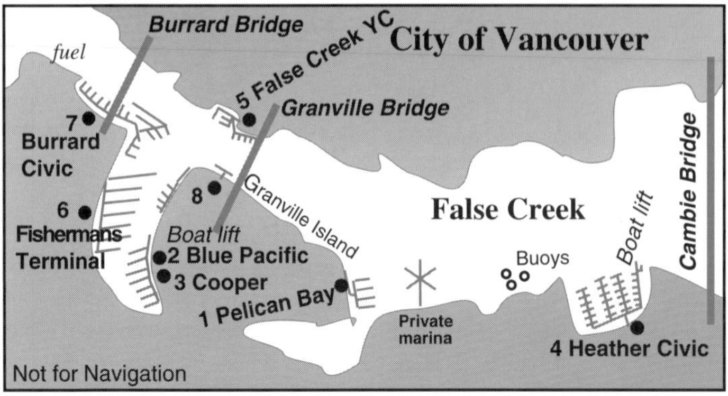

1. Pelican Bay Marina
2.Blue Pacific Charters
3. Cooper Boating Centre.
4. Heather Civic Marina
5. False Creek Yacht Club.
6. Fishermen's Terminal
7. Burrard Bridge Civic Marina (Permanent Moorage).
8. Granville Island Public dock.
Note: Manmade Granville Island is not a true island.

1. Pelican Bay Marina

Dock manager
1253 Johnston ,Vancouver, BC. V6H 3R9
Phone: (604) 682-7454 Fax (604) 683-3444

Chart 3311 False Creek

Moorage, **water, power. Showers, laundry, washrooms**. Public phones. Hotel facilities available to moorage customers. Dining. Spa. TV cable, phone hookup. (Hotel 683-7373) Boat lift and all services available on Granville Island (685-6924) 15 ton capacity.

2. Blue Pacific Charters

Phone: (604) 682-2161 Fax: (604) 682-2722

Chart 3311 False Creek

Moorage, **water, power. Showers.** Garbage disposal. Nearby remarket, restaurants.

3. Cooper Boating Centre

Phone: (604) 687-4110 Fax: (604) 687-3267

Chart 3311 False Creek

Moorage, **water, power. Showers.** Garbage disposal. Nearby market, restaurants.

4. Heather Civic Marina

600 Stamps Landing,
Vancouver, BC.V5Z 3Z1
Phone (604) 874-2814 Fax: (604) 874-6015

Chart 3311 False Creek

Moorage (mostly permanent), **water, power. Showers, laundry, washrooms.** Public phones. Pub adjacent, restaurants nearby. City shops and facilities nearby. Park, waterfront paths adjacent.
Boat lift at marina.

5. False Creek Yacht Club

Phone: (604) 682-3292 Fax: (604) 682-3614

Chart 3311 False Creek

Moorage, **water, power. Showers.** Garbage disposal. Nearby recreational centre.

6. Fishermen's Terminal

Phone: (604) 666-0180 and fax.

Chart 3311 False Creek

Moorage, **water, power. Showers**. Garbage disposal.

In Vancouver you have the alternative of an overnight in False Creek or Burrard Inlet. False Creek's Pelican Bay has facilities including hotel, spa and dining, phones and cable TV at your boat. Near the entrance of False Creek is the False Creek Yacht Club and opposite is Granville Island with several overnight options and where you will be able to access the fresh produce market, restaurants and pubs. Mooring buoys east of Granville Island are designated for use only by visiting offshore cruising yachts.

In Burrard Inlet the marinas are obvious and easy to locate after rounding Brockton Point and passing the fuel barges. Their location will give you the opportunity to visit downtown Vancouver within easy walking distance. If you wish to be away from the city, continue down Burrard Inlet and find your way to Indian Arm or Port Moody. If you prefer to anchor out overnight, go up Indian Arm to Bedwell Bay and drop anchor. Fuel is available at Seycove Marina in Deep Cove.

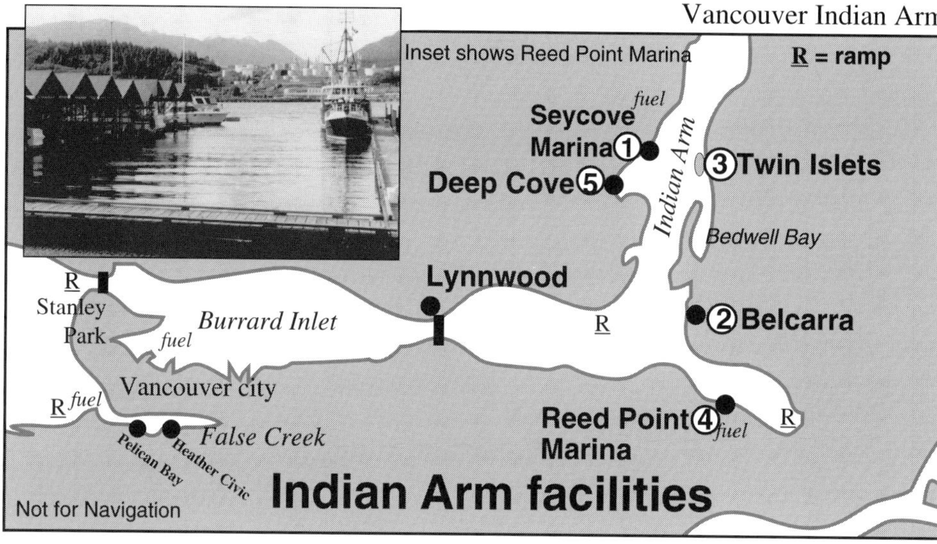

Indian Arm

1. Seycove Marina

K.J. Kim
2890 Panorama Drive, North Vancouver, BC. V7G 1V6
Phone (604) 929-1251

Chart 3495, 3435, 3311

Moorage to 70 feet, **water, power** 15 amp. **Fuel**: gas, diesel. **Washrooms.** Public phones. Chandlery, mobile repairs, service. Nearby Deep Cove village, restaurants.

2. Belcarra Park

Burrard Inlet Indian Arm
Parks Board park and floats.
Small dock mostly used for recreational fishing. Kiosk serves snacks, light meals.

3. Twin Islands

Indian Arm
Boat dock (dinghy).
Picnic sites, water, toilets, hiking, fishing, kayaking, scuba.

4. Reed Point Marina

Dock manager
850, Barnet Highway,
Port Moody, BC, V6H 1V6
Phone (604) 931-2477
Chart 3311, 3435
Marina services:
Fuel: Gas, diesel. Stove oil. Ice, water. Snacks. Bait, fishing licences. Overnight moorage, power 20 amps. 30 ton travel lift, Summer moorage for trailerable boats. Dry storage and work yard. Hamburger stand. Public phone, washrooms. Marine supplies, service repairs.
Yacht sales and service.

5. Deep Cove

Burrard Inlet Indian Arm
Transport Canada dock
Phone 757-9331
Chart 3311, 3495
Manager • Float length 44 m
Lights • Small village of Deep Cove has restaurants, groceries and other services.
Adjacent: Deep Cove Yacht Club.

The quaint, old general store at Gambier. It is located not far from the dock at New Brighton.

Howe Sound

Bracketted between Horseshoe Bay and Gibsons Landing, a score of coves, bays and anchorages welcome the yachtsman on Bowen, Gambier and Keats Islands. All have absorbing histories, but the main attractions today are astounding scenery, nearby fishing, and the quick escape from the city.

Cruising to Howe Sound for the week-end is highly possible for boat owners of almost any sized vessel moored in and around Vancouver, regardless of most weather conditions. The well-known "Squamish" wind which rips down the Sound at times ought not to be overlooked. Shelter from the Squamish is best found in lee anchorages such as Centre Bay on Gambier Island or Snug Cove on Bowen.

Going ashore at Snug Cove is like stepping onto a welcome mat at the door of Bowen Island. Land speculation seems to be the greatest industry, along with tourism, on Bowen Island. At one time a certain amount of whaling was done in the vicinity, especially in the waters around Worlcombe Island. The earliest settlers on Bowen were loggers but it didn't take long before they were homesteading and others were beginning to trek to it for summer weekend vacations and later day tripping aboard the Union Steamships for picnics, or pouring in for a weekend of swimming, dancing, hiking and a variety of sports activities.

Early this century and right up until boating came within the reach of most people, hordes

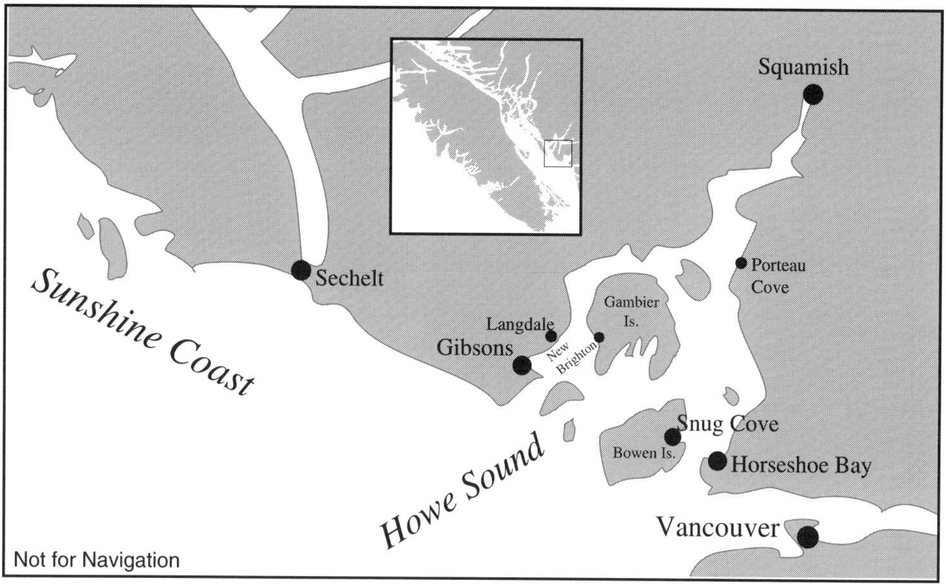

of holiday makers used the steamships leaving Vancouver at the foot of Columbia for Snug Cove on Saturdays. Eventually this "group" activity faded out as people took to doing their own things with their individual pleasure boats.

Union Steamships had put up a hotel, the Bowen Island Inn, which offered "magnificent vistas of mountains and sea." It offered visitors full course luncheons and dinners served in the "Sea-View" dining room. Moonlight cruises were offered aboard the *Lady Alexandra* and dancing aboard that vessel and at the Bowen Pavilion, were all part of the fare. A brochure put out by Union Steamships, in appealing to the public to make capital of the island's natural and other attractions, aptly describes Bowen's appeal: ". . . Lovely walks amidst glens and streams . . . through woodland trails that will take you to many beauty spots such as Lake Killarney, Trout Lake and Mount Gardner Park. On top of all this you get the restful one hour cruise each way . . . around Point Atkinson into Howe Sound, a typical British Columbia fjord of incomparable beauty, . . . towering mountains and wide sweep of sheltered sea."

A cruise around the lower reaches of Howe Sound could easily end at Snug Cove, even if one has designated that only as a first stop on a weekend's cruising itinerary encompassing Bowen, Keats and Pasley for example. Years ago I wrote:

We strolled along the Deep Bay road and forgot about time. The sun had broken through the thin veil of cloud and shone brightly on the peaks of Mount Strachan and Black Mountain looming over Horseshoe Bay opposite. It was easy, looking at the holiday and permanent cottages nestling among the trees and lush vegetation, to understand what motivated a deep interest among early as well as present day visitors and settlers to Bowen Island.

For the visiting boatsman, Snug Cove is a shelter in most weather although it can become a little uncomfortable in a bit of an easterly winter blow. For specific information on Snug Cove and other facilities at Bowen Island refer to the British Columbia Small Craft Guide. For those who like anchoring overnight caution should be taken not to drop the hook over submarine cables, nor to obstruct the path of the Bowen Island ferry.

If you want some good reading on the history of Bowen Island and the people who have come and gone and given places on the Island their names get a copy of Irene Howard's *Bowen Island 1872 -1972.* She has drawn in it a fascinating word picture of the island's past and present and through it one is more easily able to relate to the rest of lower Howe Sound, its islands and people.

Gibsons Landing and Keats Island are major boating attractions on the western side of the Sound, especially for those who like to get together with other boat owners or go ashore and either picnic or camp at the Marine Park in Plumper Cove or visit the small town stores and facilities at Gibsons.

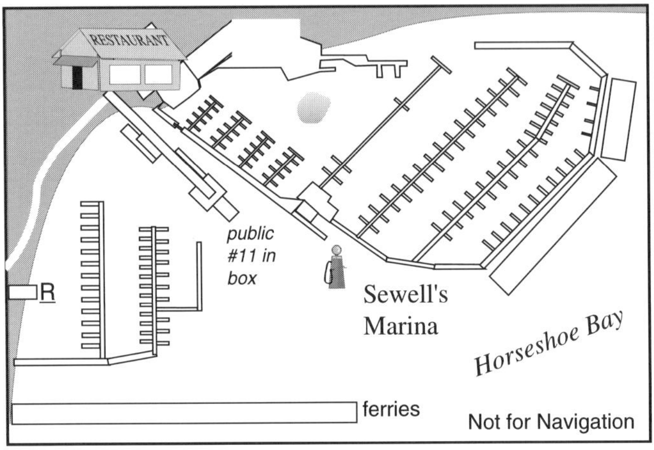

From a sleepy, quaint town with occasional coming and going of the ferries, Horseshoe Bay has turned into a busy hub of travelling activity. Sewell's Marina, top, and the ferry to Bowen Island landing at Snug Cove, above.

As a marine park, Plumper Cove on the west side of Keats attracts large numbers of boats each sunny weekend in summer. Its picnic and camping grounds are made much use of and its shady walks in the woods are popular. Views from the cliffs adjacent to the picnic and camping ground are magnificent and one can sit for hours just watching the boats cruise by or take in a colourful sunset over the opposite shore.

A cruise down Collingwood Channel could include a temporary stop at Mount Gardner Park, where there is a government float and some protection from the weather in the lee of Hutt Island.

Bowen Bay and Tunstall Bay are the large indentations on the west side of Bowen. Here it is possible to anchor and row ashore for some beachcombing. Tunstall Bay was the site in the early days of an explosives factory and the creek which runs into the bay appropriately bears the name Explosives Creek.

Horseshoe Bay

Chart 3534, 3311, 3481

11. Horseshoe Bay–map 146
West Vancouver
Transport Canada dock
Chart 3534, 3526, 3311, 3512
Manager • Float length 64 m
Lights • Power •
Adjacent restaurants, shops, ferries to Vancouver Island and the Sunshine Coast via Langdale/Gibsons.

Sewell's Marina

Derek Rendle
Dan Sewell
6695 Nelson, West Vancouver, BC.
V7W 2B2
Phone: (604) 921-3474

Marina Services

Moorage
Fuel: gas, diesel, stove oil, engine and outboard oils.
Water
Power–15, 20, 30 amp.
Washrooms nearby. Public phones, launch ramp. Chandlery, repairs, service, fishing charters, licences, bait, tackle, rentals. Restaurants, Horseshoe Bay village adjacent–many services available.

The BC Ferries docks are located alongside the large Sewell's Marina. In summer the entire, small bay is busy with the coming and going of small craft and large ferries. Rental boats operate in the midst of this activity so exercise great caution when navigating in and out of Horseshoe Bay.

Snug Cove

Union Steamship Co. Marina–Snug Cove

Rondy and Dorothy Dike
P.O. Box 250
Bowen Island, BC, V0N 1G0
Phone: (604) 947-0707 Fax: (604) 947-0708
Chart 3534, 3311, 3526, 3481, 3512
VHF call 68
email: *ussc@direct.ca*
Web: *www.steamship-marina.bc.ca*
Hazard: Watch for ferry operations. Shallows near beach beyond marina docks.

Marina services:

Moorage: 170 slips. Maximum to 220 feet.
Water at dock.
Power at docks: 50, 30 amp.
Showers, public phones, washrooms.

Customer services:

Chandlery. Boating supplies. Gifts, novelties, charts, snacks. Fresh produce, frozen foods, pharmacy, liquor, tackle, bait, hardware, ice and most supplies–at marina or in village. Restaurants, pubs, art and crafts stores, bistro, bakeries, health food and gift and specialty stores.

Entertainment.

Annual summer events include live entertainment, (pub Saturday nights),
Bowen Island parade and festival (Saturday prior to Labour Day). Dog Days of Summer (Second Sunday in August).
Walking roads and trails on island. Crippen Regional Park has many trails and walks. Including 600 acres surrounding Snug Cove and marina.

Adjacent facilities:

BC Ferries to Horseshoe Bay. Crippen Regional Park. Picnic ashore. Walking trails. Killarney Lake in Crippen Park area. Anchorage in Mannion Bay (temporary).

Balladeer log 1984–*The old general store building was being shifted from its street front location to a site a short distance back from the road. It seemed some development is being planned for the settlement of Snug Cove. On a previous visit we had gone inside and inspected the sparse array of goods on the shelves, squeaked across a creaking wooden floor and chatted to the storekeeper.* Soon after, the store was closed to make way for change. Today the building serves as a post office and offices for the Greater Vancouver Regional District. The library was housed in it for some years but was moved to the main street in Snug Cove.

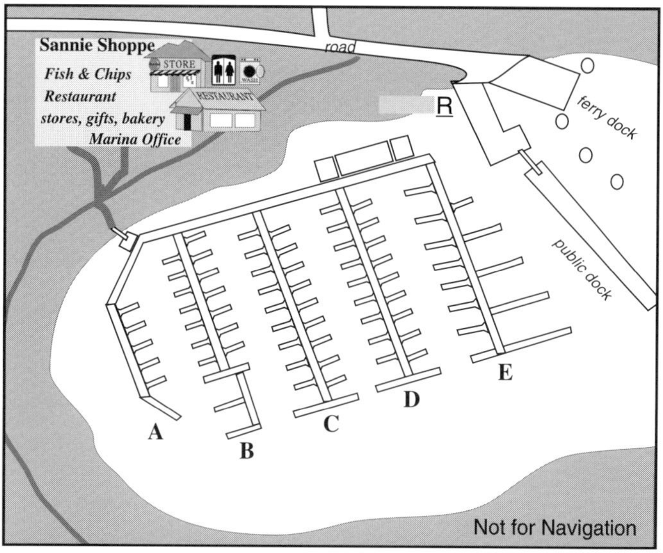

The village

Fast becoming one of the busiest boating centres on the coast, Snug Cove has been developed into a high profile marina and boating haven. Its multi slip dock facility with some 400 slips catering to boats up to well beyond 100 feet in length. The government dock that serves as a breakwater will dock about ten more boats depending on size. The ferry from Horseshoe Bay lands right alongside the government dock. Its wake and wash from the substantial propellers causes a tide-like stream that washes against boats moored to the dock so caution should be exercised when coming or going. When tying up–be sure to secure your boat adequately before leaving it for a walk up to the stores or the park and lake nearby.

The walk through the park is an easy one with well worn pathways as is the stroll up to Killarney Lake. Hiking around the lake is also not too taxing, unless you choose to do so in the middle of summer. Lots of shade helps reduce the discomfort in summertime and the cooler weather out of season does the same. Follow the trail up through the park, along the paved road a short distance to the right and then pick it up again for the stroll to the lake.

The village at the ferry landing is simply known as Snug Cove. Here there are several gift, handicraft and souvenir stores (a favourite of mine is The Sannie Shoppe) as well as refreshment establishments which include the Whirling Dervish, a light meals and coffee shop, excellent cappuccino and latte. A bakery next door serves coffee to go along with any of the freshly baked bread and pastries available. The long-established restaurant at the lower end of the stores has been serving a variety of simple meals for many years and the newer restaurant at the main Union Steamship landing boardwalk now offers a wide selection at mealtimes.

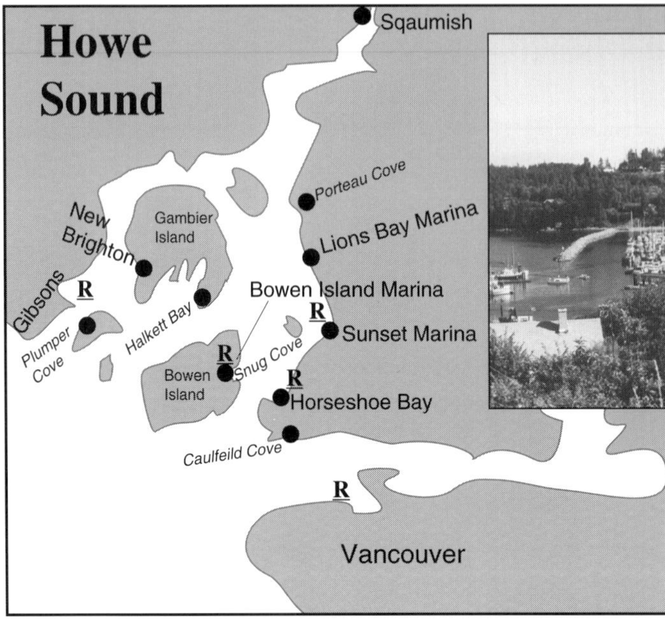

The town of Gibsons has a large marina protected by a substantial breakwater.

Lions Bay Marina

Ken Wolder
Lions Bay, West Vancouver,
BC. V0N 2E0
Phone: (604) 921-7510

Chart 3311, 3526, 3586

Moorage, water, power at fuel dock.
Fuel: gas, engine and outboard oils. Washrooms. Public phones. Launch ramp, store–marine supplies, ice, snacks, fishing licences, bait.
Closed Tuesdays and Thursdays

Sunset Marina

Sue Rauter
34 Sunset Beach, West Vancouver, BC. V7W 2T7
Phone: (604) 921-7476 Fax (604) 921-7477

Chart 3311, 3526, 3586

Wet and dry moorage, water, power at gas dock. **Launch Ramp, marine store.**
Fuel: gas, engine and outboard oils. Tackle, fishing licences. Washrooms. Public phones. Repairs, service, power wash. Tackle, bait, ice. Restaurant. Parking.

Caulfeild Cove–map 146
West Vancouver Fisheries & Oceans Chart 3311, 3526, 3463 Float length 16 m Lights • Garbage • Water •

Bowen Island Marina

Norma and Dennis Dallas
RR #1, A-1, Bowen Island,
BC V0N 1G0
Phone: (604) 947-9710

Chart 3311, 3526, 3586, 3534 VHF 16

Marina located to starboard on approaches to Snug Cove, just before ferry landing. Limited transient moorage. Reservations. Moor alongside main dock (marked).
Fishing licences, bait, tackle.
Hazard: Ferry dock alongside. Watch operations and propeller turbulence from vessels while at dock.

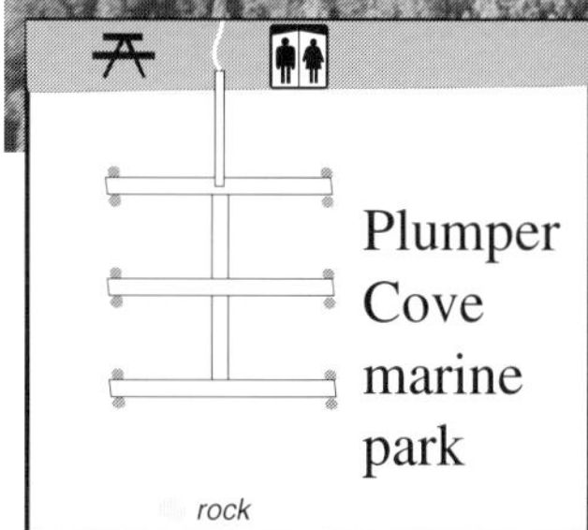

The marine park at Keats Island includes the Plumper Cove docks (below), seen beyond the ridge in the lower right of the photograph. The islands provide shelter from most conditions but occasional westerlies blow in causing some discomfort when anchored or tied to mooring buoys.

Parks in Howe Sound

Halkett Bay
All weather anchorage
(some wind conditions).
Dinghy dock,
camping sites, toilets, hiking.

Porteau Cove
Temporary moorage.
Buoys mark location of artificial reefs.
camping/picnic sites, water, toilets, beach,
scuba diving (wrecks as artificial reefs).

For more information on marine parks see
Anchorages and Marine Parks
(Seagraphic Publications 1998)

Plumper Cove
All weather anchorage
(some wind conditions).
8 mooring buoys. **Boat docks,**
camping/picnic sites, water, toilets, beach.
hiking. Fishing nearby at the Cut.

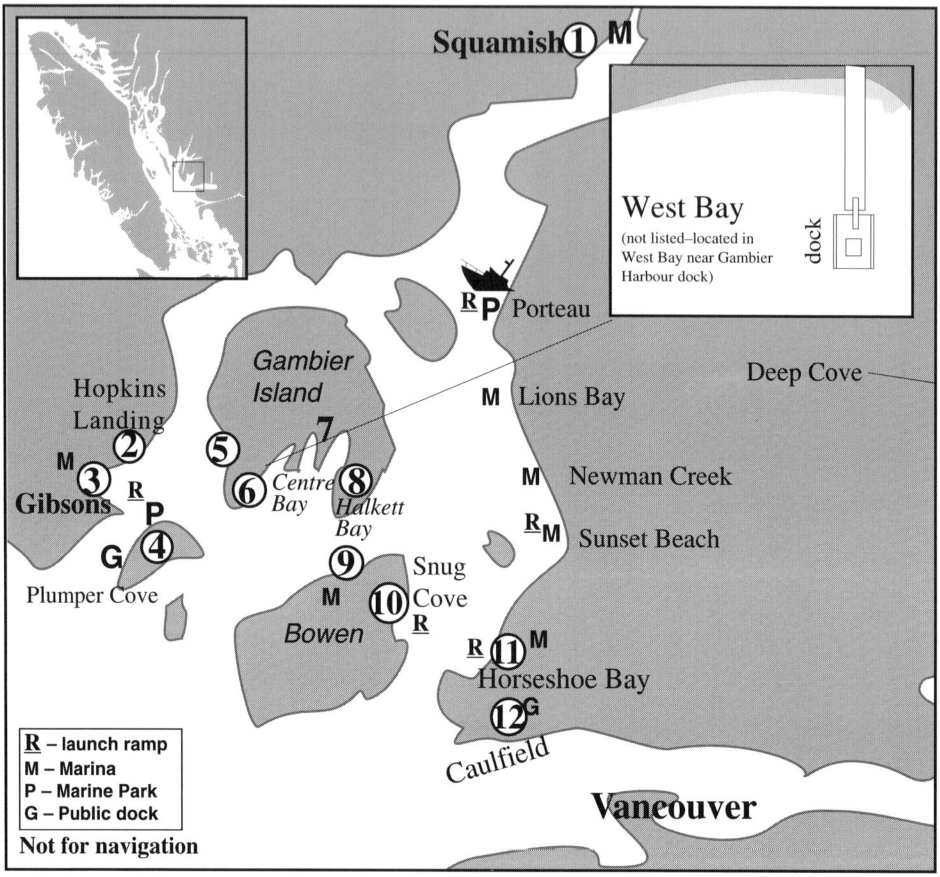

Howe Sound Public Docks

1. Squamish

Fisheries & Oceans dock
Phone: (604) 892-3908
Chart 3311, 3534, 3526
Manager • Float length 118 m
Garbage • Water • Lights • Power • Public phone ashore • Near city restaurants, shops, churches, services and facilities. Yacht club may allow some transient moorage.

2. Hopkins Landing

Transport Canada dock
Chart 3311, 3526, 3512
Manager • Float length 17 m
Lights •

3. Gibsons

Fisheries & Oceans dock
Phone: (604)886-8017
Chart 3534, 3311, 3526, 3512
Manager • Float length 407 m
Launch ramp • Breakwater • Aircraft float • Garbage • Waste oil disposal • Water • Lights • Power • Public phone ashore • Nearby restaurants, shops. Gibson's Marina adjacent.

4. Keats Island

Transport Canada dock
Chart 3311, 3526, 3512
Manager • Float length 15 m
Water • Lights •

The public dock at New Brighton is a favourite stop for a light snack, groceries or an ice cream cone up at the old general store (page 138).

5. New Brighton

Gambier Island
Transport Canada dock
Chart 3311, 3526, 3512
Float length 120 m •
Lights • Public phone •
Walking–island roads.
Nearby store.
Ferry service to Langdale/
Gibsons.

6. Gambier Harbour

Gambier Island
Transport Canada dock
Chart 3311, 3526, 3512
Float length 30 m
Lights • Showers •

7. Port Graves

Gambier Island
Transport Canada dock
Chart 3310, 3512, 3526
Manager • Float length 10 m

8. Halkett Bay

Gambier Island
Transport Canada dock
Chart 3311, 3526, 3512
Float length 17 m
Adjacent marine park

9. Mount Gardner Park

Bowen Island
Fisheries & Oceans dock
Chart 3311, 3526, 3512
Float length 17 m

10. Snug Cove

Bowen Island
Transport Canada dock
Chart 3534, 3311, 3481, 3512, 3526
Manager •
Float length 105 m
Garbage • Lights • Power •
Public phone ashore •
Washrooms ashore •
Adjacent Union Steamship Marina.
Near village arts and crafts, bakeries, restaurants, shops.
All services.
Ferry to Horseshoe Bay.

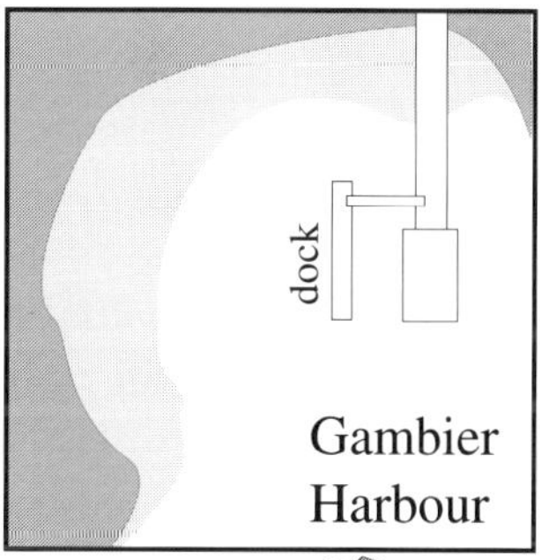

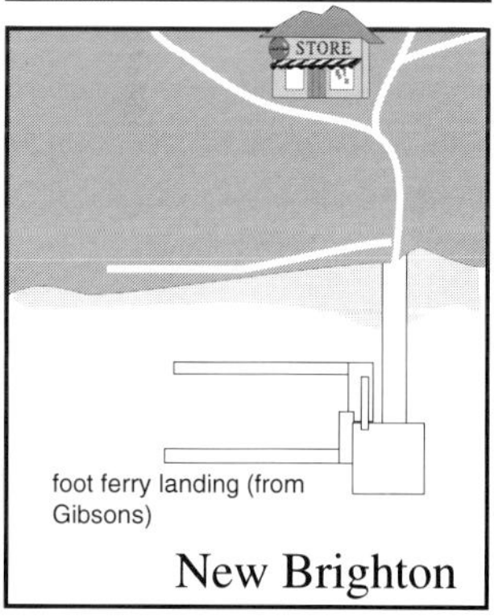

Not for Navigation

Gibsons. Transport Canada dock
(604) 886-8017 Manager • Float length 804 m
Breakwater • Water • Lights • Power •
Public phone ashore • Near restaurants, shops.

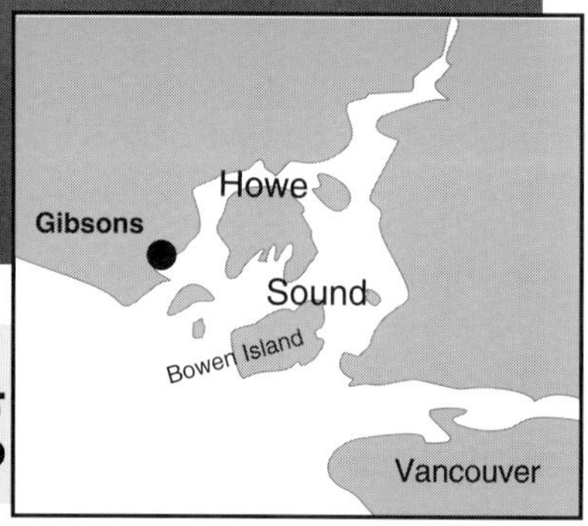

Gibsons Landing

Gibsons Marina

Art McGinnis
P.O. Box 1520, Gibsons,
BC, V0N 1V0
Phone: (604) 886-8686
Fax: (604) 886-8686
Charts 3534, 3311, 3526, 3512 VHF 68

Marina services:

400 permanent and transient berths. Dock A – main visitor dock.
Water at dock. Many outlets.
Sani-station (pumpout).
Power at docks: 15 amp at all slips.

Customer services:

Laundry, showers, washrooms.
Marine supplies at chandlery/marina office. Fishing gear, licences, charts, bait, ice, repair accessories, books, binoculars. Nearby church/es multi-denominational.
Post Office: Walk to stores in village, also pharmacy and other necessities.
Scuba diving at Sechelt up the Sunshine coast. Ask for information.
Public pay phone ashore.
Walking: Road access walking or cycling. Waterfront walk, partial around bay. Walk, cab or bus up to main shopping centre.
Hourly bus service between Langdale ferry, Gibsons and Sechelt.

Entertainment.

Fishing excellent at The Cut a short run from the marina.
Museum, local stores. Art galleries.

Adjacent facilities:

Fuel: Gas, diesel, oils. Outboard mix.
Service, haul-outs.
Marine repairs available–mechanic on call. Marine and boat equipment sales. Bank machines, gifts, arts and crafts, snacks. Vehicle rentals. **Launch ramp.**

The floats at Gibsons Marina are higher than average, sturdy and well-equipped with water and power.

Right: A view of Gibsons Landing government docks from Smitty's Marina (in the foreground). Gibsons Marina in the background.

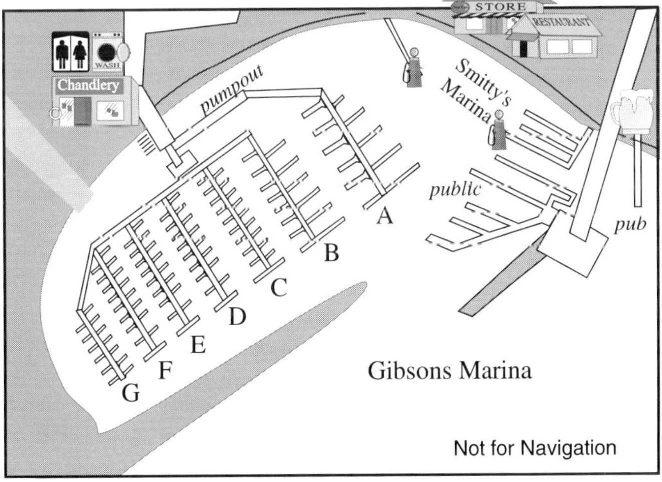

Gibsons to Squamish

Gibsons today is a busy community of residents, transient ferry travellers, business people and boat operators docking their craft at Gibsons Marina. This large facility is one of the biggest and best of its kind in local coastal waters. It provides all services required by mariners from the chandlery store at the head of the dock to nearby repair facilities. The marina offers power and water, showers and laundry, fish cleaning, bait, tackle and accessories. In the waterfront village of Gibsons Landing there are stores, restaurants, a

Squamish

Moorage in Squamish

The Squamish Yacht Club has moorage alongside the public marina. There is often adequate room to tie up at the public docks and sometimes a yacht club member will invite you to stop for a while at their facilities, provided there is room. Many mariners tie up to log booms in the area but be careful not to hinder any work in progress if you do. Stop and enquire at the Yacht Club if the government docks are full.

Photos: Views from the public docks alongside the Squamish Yacht Club.

pub, post office, pharmacy, museum and supermarket. A very handy delicatessen is located near the marina for excellent specialty items and a sit down cup of tea or coffee. Craft and art stores have become a major attraction.

From Gibsons to Squamish takes you either up the eastern shore of Howe Sound, where you can see numerous waterfront homes or small settlements, or up the western side with its islands and passages. One of the major landmarks on the steep-sloped eastern shore is Britannia Beach with its scarred hillside from former mining operations.

The passage up the western shore passes from Gibsons via New Brighton on Gambier, privately owned club facilities of Thunderbird Yacht Club and Burrard Yacht Club at Ekins Point and McNab Creek opposite.

Sunshine Coast

Howe Sound to Desolation Sound

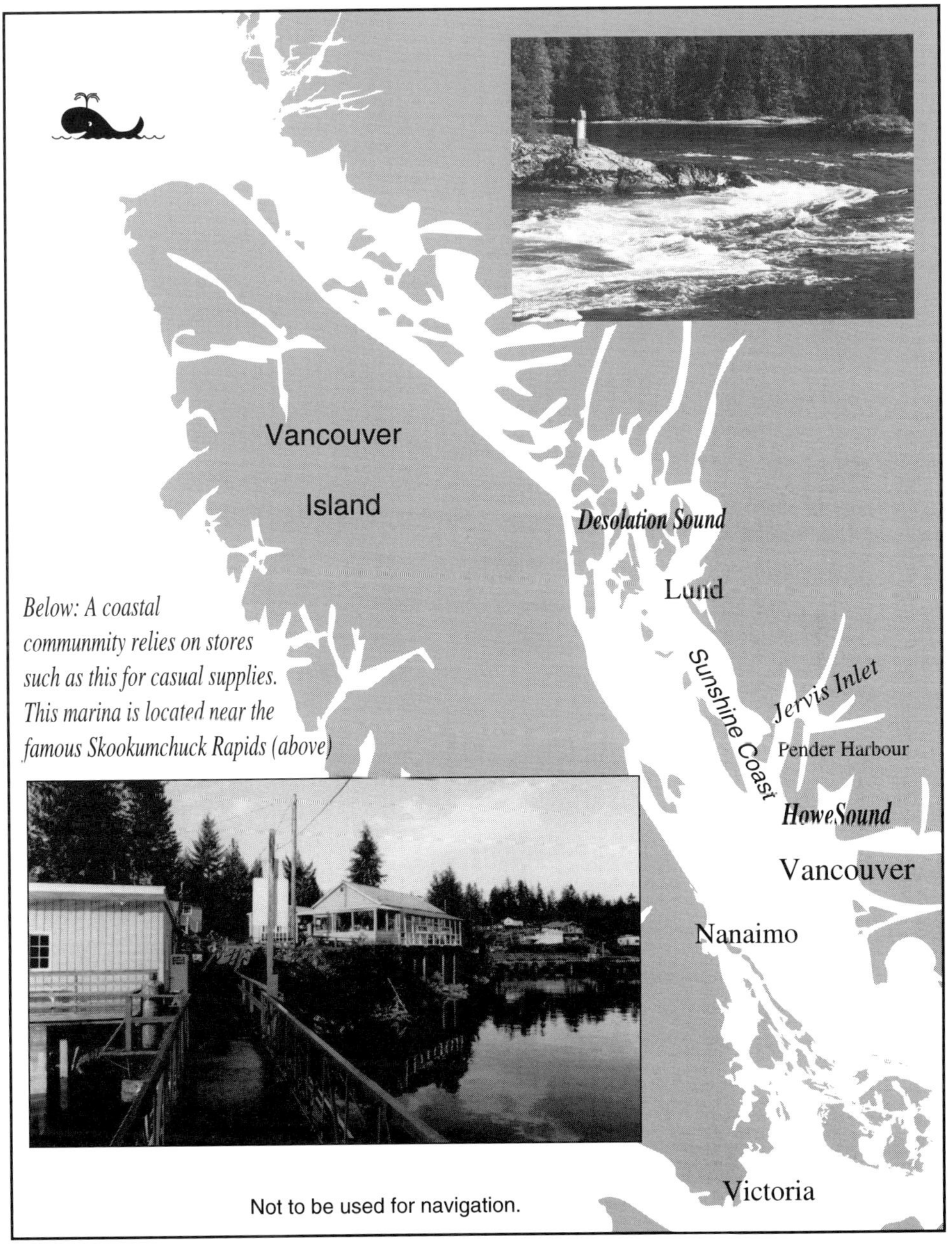

Below: A coastal communmity relies on stores such as this for casual supplies. This marina is located near the famous Skookumchuck Rapids (above)

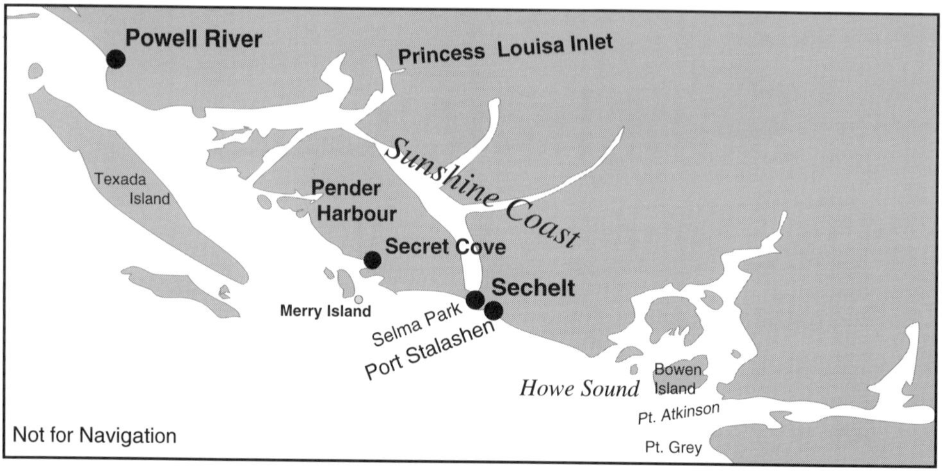

Above: A rough water refuge can be found between Secret Cove and Howe Sound at Selma Park just south of Sechelt on Indian reserve or at Port Stalashen (below) on the approach to Trail Bay from Gibsons.

Port Stalashen Marina

Art McGinnis *(see also Gibsons Marina)*
RR #1 Field Site C26, Sechelt
BC, V0N 3A0
Phone: (604) 885-4884 Fax: (604) 885-3101
Charts 3311, 3512. VHF 68

Marina services: Some guest moorage.

Water. Power: 30 amps. Pumpout.

Customer services:

Showers, washrooms. Garbage.

Public pay phone ashore.

Nearby: Restaurants, accommodation.

Use chart 3311 and enter with great care during rough conditions.

Port Stalashen makes a good stop en route off Sechelt when the weather becomes unsettled.

The approximate dock layout has been superimposed over this photograph, taken while the marina was under construction.

Secret Cove

Secret Cove Marina

Dock Manager
Box 1118 Sechelt,
BC, V0N 3A0
Phone: (604) 885-6037
Chart 3535, 3311, 3512

Marina services:

Fuel: gas, diesel, oil, live bait, ice,
150 berth permanent and transient docks.
Water at dock. Many outlets.
Power at docks: 30, 15 amp.

Customer services:

Showers. Washrooms. Fish cleaning tables, picnic tables, garbage disposal for moorage customers only. Marine supplies at chandlery. Fishing gear, licences, charts, bait, ice, electronics, repair accessories, books, gifts, arts and crafts, snacks. Fax service.
General store. Public pay phone ashore.
Water Taxi available.
Walking: Road access walking or cycling. Cab or bus up to main shopping centres.

Entertainment.

Good fishing a short run from marina.
Hotels, restaurants and facilities nearby.

Adjacent facilities:

Golf. Swimming

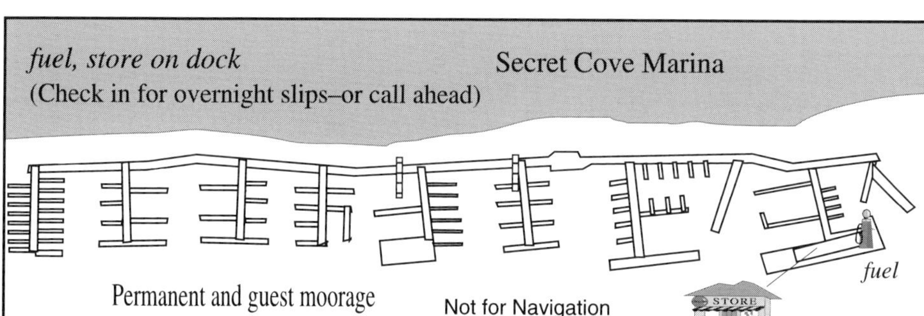

Buccaneer Marina

Bob, John, Gerry Mercer
RR 1 Halfmoon Bay
BC, V1Y0
Ph (604) 885-7888 Fax: (604) 885-7824
Chart 3311, 3535, 3512 VHF 68

Marina services:

Haulouts. **Fuel:** gas, diesel, oil, live and frozen bait. Marine centre and store. Sales & service. Parts and repairs. Mercury Marine service. Permanent moorage and repair dock. **Water** and **power** at dock.

Customer services:

Boat rentals. Fishing gear, licences, charts, bait, ice, propane, groceries, snacks. Hull repairs, steam cleaning, bottom painting. Public pay phone ashore.
Walking: Road access walking.

Secret Cove,

Fisheries & Oceans dock
Chart 3311, 3535, 3512 Manager • Float length 44 m • Lights • Power • Public phone •
Adjacent marinas, lodges, provisions, fuel, repairs, service.

Adjacent facilities:

Moorage, Secret Cove Marina. Accommodations, Lord Jim's Lodge (604) 885-7038, and Jolly Roger Inn (604) 885-7184. (Limited overnight guest moorage.)

Porpoise Bay, Sechelt

Fisheries & Oceans dock
Phone 885-1986
Chart 3311, 3512
Manager • Float length 132 m• Launch ramp • Grid • Aircraft float • Garbage • Water • Lights • Power • Public phone ashore • Adjacent restaurant, hotels, nearby shops. Scuba diving in vicinity. Charters. Rentals.

Halfmoon Bay,

Sunshine Coast
Transport Canada dock
Chart 3311, 3512
Float length 26 m • Lights •

Sechelt,

Sunshine Coast
Fisheries & Oceans dock
Chart 3311, 3512
Manager • Breakwater •

Sandy Beaches, Sheltered Coves

The focal point on the Sunshine Coast on the Vancouver side of Pender Harbour is Secret Cove with nearby Smuggler Cove and Buccaneer Bay. It's a total distance of about 40 miles from Vancouver so that even with a slow boat it is an easy run. Ideal for a day's fishing in the vicinity, especially just north of Thormanby Island and at Tottenham Ledge — most popular fishing grounds for local anglers, and visitors too.

The entrance to Secret Cove lies east of the entrance to Smuggler Cove. Parts of the west side of the inlet dry at low tide, although they looks deceivingly navigable at high tide. The lee of Turnagain offers sheltered anchorage for those who prefer to remain untethered during the night. Secret Cove measures about a mile in length and is about a quarter of a mile wide with two navigable arms at the eastern side. Secret Cove Marina with its new extensions and its well-stocked floating store and fuel station is a busy place in the summer. The marina has a lot to offer and many docks for overnight moorage. The store on the docks sells groceries, fishing tackle and most items that cruising people look for when they are out boating.

There are private floats in the southern arm and Buccaneer Marina occupies a significant area in the other. At this marina, owned and run by the Mercer family, repairs and servicing are available as well as gas, diesel fuel and small boat launching (for those who choose to arrive by road the marina operators will put trailerables in the water with a four wheel drive). Many property owners on Thormanby Island across Welcome Passage maintain boats in Buccaneer Marina permanently and park their cars on the Mercers' property when they take off to their island homes at week-ends.
Getting in and out of these arms is regarded by some as tricky, but this is not really the case if one follows a chart carefully and takes heed of the various markers.

If you have any doubt about navigation in this area a good book to use is the B. C. *Small Craft Guide.*

The Jolly Roger Inn and Lord Jim's Lodge at Secret Cove offer hotel accommodations, restaurant and other luxury facilities for those who need a break from boating.

The Union Steamship ferry that plied the coast from Vancouver to Pender Harbour used to call at many stops on the way and these included Halfmoon Bay and Sechelt. The boats that called in Secret Cove, however, were mainly commercial or occasional pleasure boats— sometimes carrying groups of weekending holidaymakers, most of whom made their destination the golden sands of nearby Buccaneer Bay.

Romantic as the name "Buccaneer Bay" may sound it does not stem from any historic facts concerning pirates and buried treasure, but rather from a race horse in the 1860 Derby held at Epsom in England. The name was given by Captain Richards who was surveying the area at the time and it appears he and his men were quite influenced by the Derby because other names, including Merry Island, Thormanby, Welcome Passage and Epsom Point have similar origins. The horse, Thormanby, belonging to J.C. Merry (Merry

Island) won the race while one of the favourites went by the name of Buccaneer.

The area was settled by immigrants from Europe and back east. Researching the history of places such as Gibsons, Roberts Creek and many others one could unearth a wealth of interesting information concerning early days on the Sechelt Peninsula. The new settlers were responsible for clearing land to build homes and church or community halls, schools and government wharves. They planted the lands and were responsible for the beginnings to the Peninsula Highway, which incidentally reached Sechelt from Gibsons and Langdale as early as 1912, Halfmoon Bay in 1930 and Pender Harbour in 1936.

Smuggler Cove hides from Welcome Passage behind the rocky breakwater that is Wilbraham Point and whose entrance is practically due south of Pender Harbour and less than a mile out of Secret Cove. Smuggler Cove (not Smuggler*s*) Marine Park is one of the best anchorages on the coast. Its narrow entrance, dotted with rocks and tiny islets, is almost deceiving to the uninitiated. Make sure you are in the right channel as you enter.

To reach Buccaneer from Smuggler Cove it is wise to travel all the way around the buoy at the northern edge of Tottenham Ledge. Down into the bay the expansive stretch of golden sand lures you on until you are in among the many other boats, sail and power, anchored and rafted within a short row of the beach.

Picnicking or just strolling on the beach as well as dipping in the cool water are the favourite ways to enjoy this area in Buccaneer Bay. The beach, which divides North Thormanby Island from South Thormanby, extends up to Vaucroft beach on the west side of the bay. There is a public wharf which provides limited berthing space in 18 feet of water. For overnight anchorage the most suitable place for small craft is on the opposite shore at Water Bay which opens off the east side of Gill Beach. Another overnight anchorage could be found in the lee of Surrey Islands just off the northern end of South Thormanby Island. Please refer to ***Anchorages and Marine Parks.***

Make certain your children are wearing life jackets around the water, and particularly on marina docks.

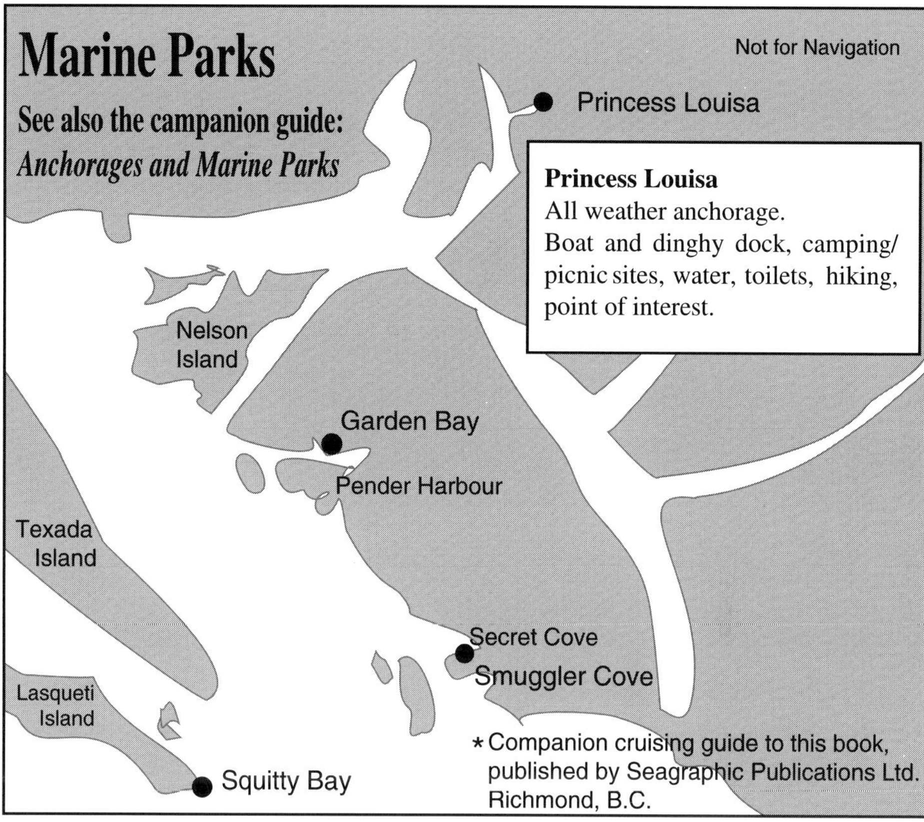

Garden Bay
All weather anchorage. Dinghy dock, (Adjacent marinas), picnic sites, toilets, park host, beach, hiking. Nearby good fishing, kayaking and scuba diving.

Smuggler Cove
All weather anchorage, camping sites, toilets, park host, hiking, kayaking in vicinity.

Squitty Bay
All weather anchorage (limited room and some wind). Boat dock, picnic sites, water, toilets. Kayaking in vicinity.

Some vessels visit Princess Louisa early in the season to capture the loneliness and tranquillity of the place. Short days and long cool nights provide a brisk but refreshing atmosphere to this unique park.

Pender Harbour

Aside from its scenic value, which is considerable, Pender Harbour has long attracted yachtsmen as a base for fishing and supplies, well away from the cities, yet readily accessible for a weekend.

When the wet days of winter are over and the sunshine of summer draws myriad boats back into the water, that new but enigmatically familiar urge for adventure spurs on the enthusiastic yachtsman to dig out his charts and begin to plan his weekend cruising.

While it is probably every boat owner's dream to put into some hitherto unknown and unexplored inlet—such a luxury is not for most weekenders out of Vancouver or Vancouver Island. For them, rather, a destination that will prove among the most appealing is bustling but strangely tranquil Pender Harbour.

The scenic attractions of Pender Harbour are many and varied along the inlet's jagged 32 miles of shoreline, but many eyes may be blinded to this by the extensive development along its shores.

The weekending boatowner heading for Pender Harbour is attracted mainly by the fishing to be done there. The waters around Pender are well-known for their yield of lingcod, salmon, and various species of rockfish. Or one can enjoy the inlet just lying at anchor in one of the many sheltered coves or putting into a marina and exploring some of the local shore-side attractions.

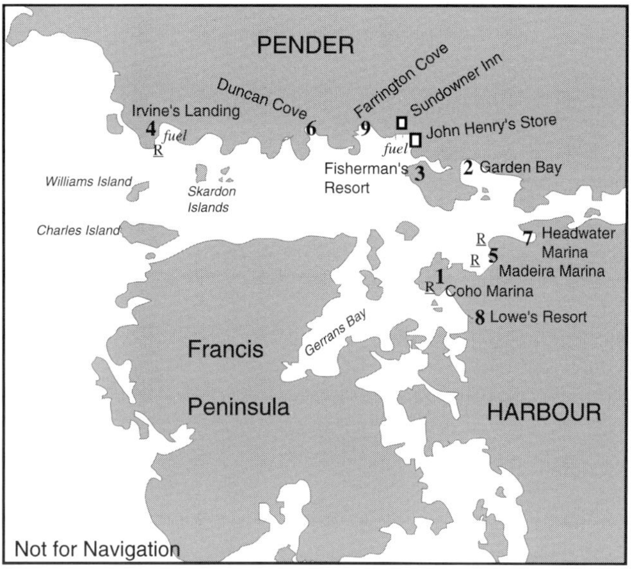

Opposite: Aerial view of Garden Bay and Hospital Bay in Pender Harbour. Garden Bay Hotel Pub marina. Carla walks down the ramp to the dock. The restaurant is seen in the background.

It's usually a quick trip in a fast power boat and on a good day it can be done in about two hours out of Vancouver. But it can also take a day if you linger over some of the engaging distractions en route, including nosing into Halfmoon Bay, one of the stops usually missed on any cruise up the Sunshine Coast. Halfmoon Bay was once the halfway stop for the Union steamships carrying tourists to Pender Harbour.

The government wharf there, however, offers limited moorage and poor shelter from any severe weather.

In 1929 the Union Steamship company vessels were doing the run to Pender Harbour from Vancouver three times a week. The trip would take anywhere from five to 10 hours—and getting back home entailed an indefinite wait on the wharf, often throughout a hot afternoon and sometimes into the evening, for the boat to arrive. In those days Irvine's Landing at the entrance to Pender Harbour was the terminal dock for travellers. From the earliest days of settlement until the 1930s when the Sechelt Highway reached the area, Irvine's Landing was one of the main centres of growth and activity.

It was about that time that other centres began springing up to cater to the many visitors and new land owners who began to pour into the area. Places such as Pope's Landing and Donnely's Landing on the Francis Peninsula and Madeira Park were established or began coming into their own. And the Union ships included them in their stops.

Entering Pender Harbour after rounding Francis Peninsula, pass between Martin Island to port and Charles and William Islands to starboard. Irvine's Landing is situated in the tiny cove called Joe Bay immediately beyond Henry Point on the northern shore of the inlet. Joe Bay takes its name from Portuguese Joe Gonsalves, who with his son-in-law, owned the first two buildings at Irvine's Landing, the hotel and the saloon.

Today the favoured overnight stops include Irvine's Landing, Fisherman's Marina, Garden Bay Hotel and Marina, Madeira Marina and Coho Marina. People stopping at the government docks at Madeira Park may walk over to Lowe's Marina for a shower or to use the other facilities such as laundry. Their marina is basically dedicated to regulars and permanent moorage and the diehard fishing enthusiasts who operate from there year round.

The one-time hospital at Hospital Bay is now a lodge and caters to dining and overnight lodging guests. Yachts tying up at marinas in Hospital Bay and Garden Bay will not be far from the facility. Coho Marina is across Pender Harbour from Hospital Bay and if a dinner is planned at Hospital Bay there should be ample dock space at the government float. Temporary tie up of a dinghy for the purpose of using adjacent restaurant or store facilities should be no problem at most marinas. Just advise the operators of your intentions and get their permission.

Pender Harbour

1. Coho Marina Resort

Gordie and Nichole Standal
P.O. Box 160 Madeira Park BC
V0N 2H0 Phone: (604) 883-2248
Chart 3512, 3535, 3514 VHF call 68
Hazard: Rocks in bay–marked by beacons. Consult charts.

Marina services:

Moorage. Launch ramp.
Water at dock. Ice. **Showers.**
Power at docks: 15 amp.

Customer services:

Fishing tackle, licences. Marine hardware.
Washrooms. Campsites, trailers.
Nearby church/es: multi-denominational.
Public pay phone ashore.
Walking: Road access walking or cycling.

Entertainment.

Kayaking. Good scuba diving and fishing a short run from harbour.

Overlooking Lowe's Marina with Coho Marina in the background. Coho Marina can be seen at left in the photo at top.

Adjacent facilities:

Stores/shopping centre, marine service.

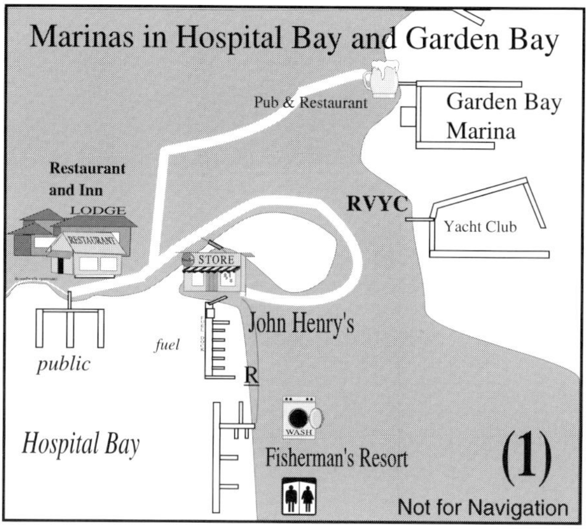

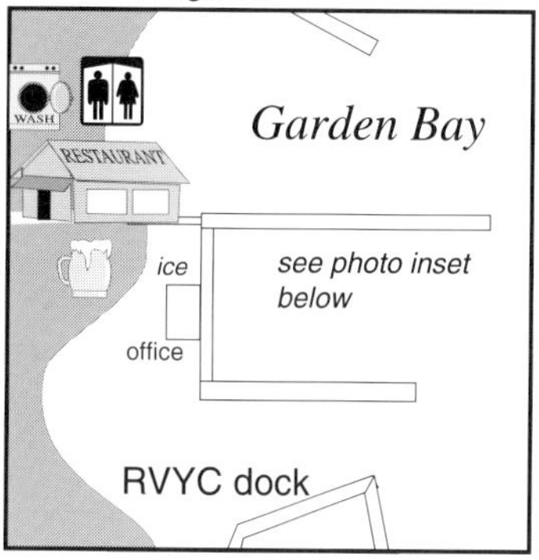

2. Garden Bay Hotel & Marina

The floats at Garden Bay Hotel & Marina are shown in the inset photograph. Garden Bay is beneath in the aerial photograph.

Dock manager
P.O. Box 90, Garden Bay,
BC, V0N 1S0 **VHF 68**
Ph: (604) 883-2674

Marina services: **Chart 3535, 3311, 3512**

Fuel: at John Henry's Marinas in Hospital Bay. Gas, diesel, stove oil, outboard fuel, propane. Tackle.

Moorage. Water at dock.

Power : 30, 15 amp.

Customer services:

Fine dining at pub, and waterfront restaurant.

Fishing charters. Air charters.

Laundry, **Washrooms.**

Gift shop and art gallery.

Nearby church/es.

Post Office: Public pay phone ashore.

Walking: Road access walking or cycling.

Entertainment.

Good fishing, and scuba diving a short run from the harbour.

Adjacent facilities:

Nearby–fuel dock (John Henry's), liquor agency, restaurants, anchorage.

3. Fisherman's Resort & Marina

Wally and Susan Nowik
Box 68 Garden Bay
British Columbia, V0N 1S0
Phone/fax: (604) 883-2336

Chart 3535, 3311, 3512 VHF call 68

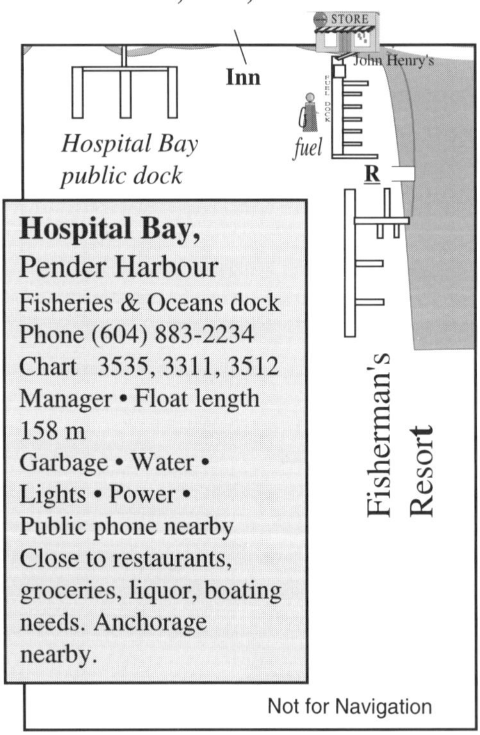

Hospital Bay,
Pender Harbour
Fisheries & Oceans dock
Phone (604) 883-2234
Chart 3535, 3311, 3512
Manager • Float length 158 m
Garbage • Water •
Lights • Power •
Public phone nearby
Close to restaurants, groceries, liquor, boating needs. Anchorage nearby.

Marina services:

Moorage. Launch ramp.
Power at docks: 30, 15 amp.
Water.

Customer services:

Laundry, showers, ice, bait.
Waterfront cottages. Public phone ashore.
Marine charts, books. Fishing gear, licences.
Nearby church/es: multi-denominational.
Scuba diving arrangements and charters –ask marina for details.
Walking: Road access walking or cycling.

Entertainment:

Fishing is excellent in general area out of Pender Harbour. Consult fishing guides.

Adjacent facilities:

Groceries at John Henry's general store. Also, post office, liquor, restaurants.

Pharmacy and other necessities at store. Sundowner Inn Restaurant. Phone: 883-9676. Public marina. Anchorage (in Garden Bay). **Fuel** at Irvine's Landing or John Henry's.

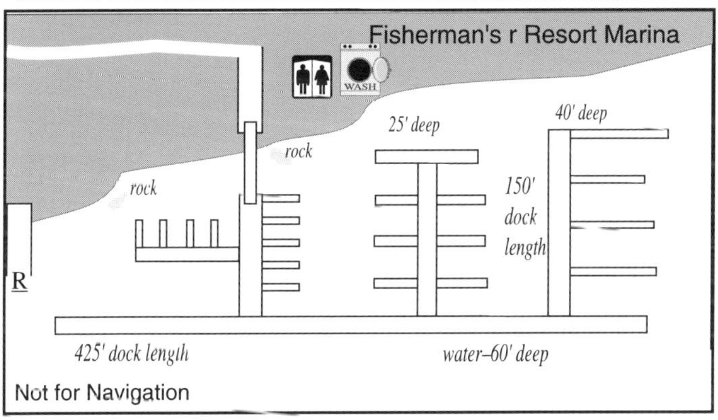

A view over the bay from Sundowner Inn includes John Henry's fuel dock in the foreground and Fisherman's Resort beyond. The marina, busy in the early part of summer.

The fuel dock at John Henry's.

Top: Approaching the dock at Fisherman's Resort.
Above: Irvine's Landing with fuel dock and space for overnight moorage.

Irvines Landing,
This was once a public dock operated by Fisheries & Oceans. It is located at the entrance to Pender Harbour and offers guests moorage, fuel, groceries and many of the needs of cruising mariners, as well as the many sports fishing enthusiasts who frequent the facility. It has a pub style restaurant and patio service.

4. Irvine's Landing

Marina & Pub

Dave or Lorna
RR #1, Irvines Landing Road,
Garden Bay, B.C. V0N 1S0
Phone: (604) 883-2296 Fax(604) 883-2080
Phone: 883-1145 (Marina & Pub)
Chart 3535, 3311, 3512

Marina services:

Fuel: gas, diesel, outboard mix.
Transient moorage.
Power. 30, 15 amp. **Water.**

Customer services:

Irvine's Landing Pub. Dining. Patio service. Wine & beer store. Groceries. Tackle, charts, live and frozen bait. **Showers, washrooms.**

Excellent fishing nearby. Guides. Fish freezing. Two fish cleaning stations. Boat rentals–guided or unguided. Marina conveniently located at entrance to harbour.

Entertainment.

Walks, hiking–parks and roads. View points. Animal and marine life.

Adjacent facilities:

Launch ramp. Newly renovated. Kayaking, fishing charters. **Pay phones.**

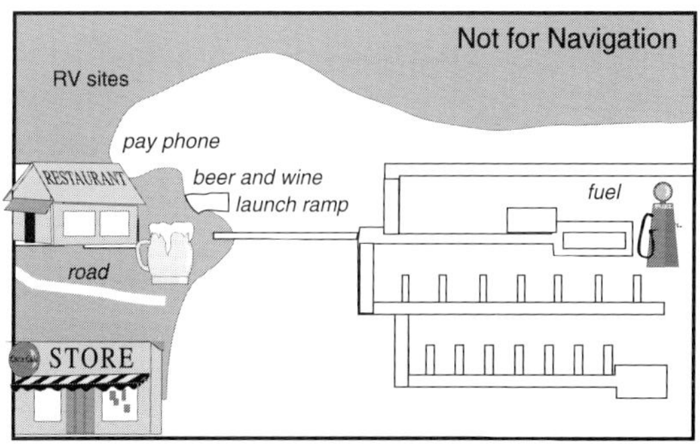

Madeira Park—Pender Harbour

5. Madeira Marina

Karen Reid and Rick King

P.O. Box 189 Madeira Park, BC. V0N 2H0

Phone: (604) 883-2266 Fax (604) 883-9250

Chart 3512, 3535, 3514

Location: next to Madeira Park public docks and ramp.

Marina services:

Moorage. For customers at lodge, and adjacent RV park, also service customers. **Water** at dock. Ice. No power at docks.

Customer services:

Marine store and supplies, fishing tackle, licences, ice, freezing.

Walking: Road access walking or cycling.

Entertainment.

Scuba diving and fishing excellent a short run from the harbour.

Adjacent facilities:

Public dock. Nearby supermarket, post office, bank, pharmacy, restaurant.

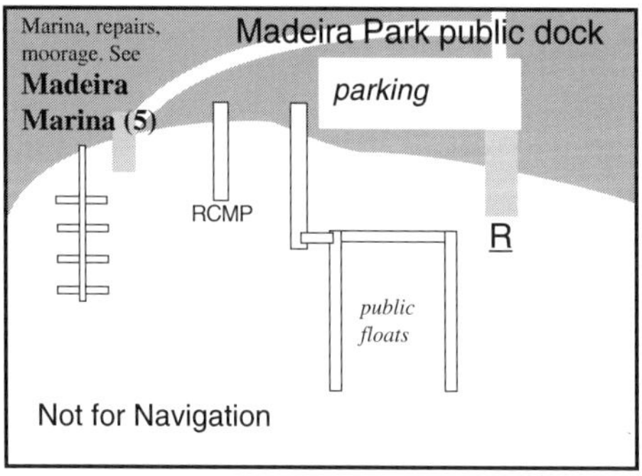

6. Duncan Cove Marina

Albert Hull

4686 Sinclair Bay Road, Madeira Park, BC. V0N 1S0

Phone: (604) 883-2424

Fax: 883-2414

Chart 3512, 3535, 3514

Marina services: Showers, laundry, washrooms. Ice, **water, power**. Groceries, fishing tackle, licences, garbage disposal, rental boats. Walking. Take care–we have seen bears on the roads.

Madeira Park,

Pender Harbour
Fisheries & Oceans dock
Phone 883-2234
Chart 3535, 3311, 3512
Manager • Float length 196 m
Launch ramp • Aircraft float • Garbage •
Waste oil disposal •
Water • Lights • Power • Public phone •
Nearby restaurants, shops.

Top: Aerial photo shows Lowe's Resort to the right (Coho Marina to the left). Above, right: Lowe's Resort Marina. Coho Marina beyond.

7. Headwater Marina

Pender Harbour
Gunboat Bay, B.C. V0N 2H0.
Phone: (604) 883-2406
Chart 3535, 3311, 3512 VHF 73

Marina services:

Transient moorage.
Power. Water.

Entertainment.

Walking on roads. Animal and marine life.

Adjacent facilities:

Kayaking, fishing charters. Anchorage nearby–consult local chart.

8. Lowe's Resorts

Davenia and Reg Morton
Box 153, Madeira Park, BC.
V0N 2H0
Phone: (604) 883-2456 Fax (604) 883-2474
Chart 3535, 3311, 3512
Moorage. Limited–primarily for lodge customers. Boats to 28/30 feet.
Manager–Larry Curtis.
Water, power.

Customer services:

Showers, washroms, tackle store, snacks. Accommodation.

Entertainment:

Boat rentals. Kayaking, fishing, scuba diving. Shops nearby.

Note: *Moorage guests at Madeira Park or other transient boat crews are welcome to use coin operated laundry and showers at this marina. Access is by way of an easy walk from Madeira Park public docks.*

Chart 3512, 3535, 3514 VHF call 68

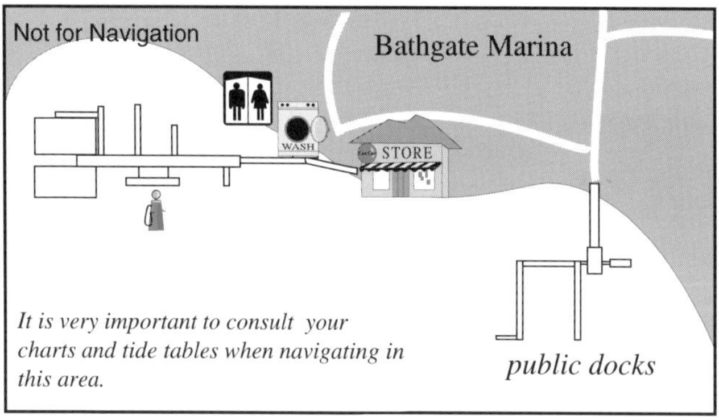

Opposite: Bathgate Marina and public docks. Inset: Bathgate Marina docks and store with laundry and showers in building on left. Above: Docks and fuel float at Bathgate Marina. Below: Public docks at Egmont.

Public launch ramp nearby at Egmont.
Also gravel ramp at Egmont Marina.

Egmont

Bathgate General Store & Marina

Doug and Vicki Martin
Egmont BC V0N 1N0
Phone: (604) 883-2222 Fax 883-2750
email: *bathgate_egmont@sunshine.net*
Web: *www.bathgate.com*

Hazard: Drying reef in middle of bay on approaches to fuel dock and another off the government dock. Check your chart.

Marina services:

Fuel: Gas, diesel, propane, oils.
Moorage. Marine ways. Marine mechanic on duty.
Water at dock. Ice.
Power at docks: 15, 20, 30 amp.

Customer services:

Liquor agency. Grocery store–fresh meat and vegetables. Fishing tackle, licences. Water taxi service.
Washrooms, showers, laundry.
Boat, cabin, video rentals.
Public pay phone ashore.
Walking: Road access walking or cycling. Nearby hiking trails. Trail to the Sechelt Rapids viewpoint in the Skookumchuck Provincial Park.

Entertainment.

Scuba diving and fishing in the immediate vicinity is rated very highly. Kayaking. Fishing and scuba diving charters.

Adjacent facilities:

Public dock. Base before trip to Princess Louisa Inlet. Automobile fuel.

Egmont Public dock

Fisheries & Oceans dock
Phone (604) 883-9243
Chart 3512, 3535, 3514.
Manager • Float length 145 m
Aircraft Float • Garbage • Waste oil service • Lights •

Egmont Marina at the Back Eddy Pub in Egmont. Opposite: Launching at the ramp adjacent to Royal Reach Marina.

Egmont Marina

John and Margaret Mills
General Delivery,
Egmont, BC V0N 1N0
Phone/fax: (604) 883-2298
Tollfree 1 800 626-0599
Chart 3311, 3512 .
email:
egmont_marina@sunshine.net
Web: *www.egmont-marina.com*

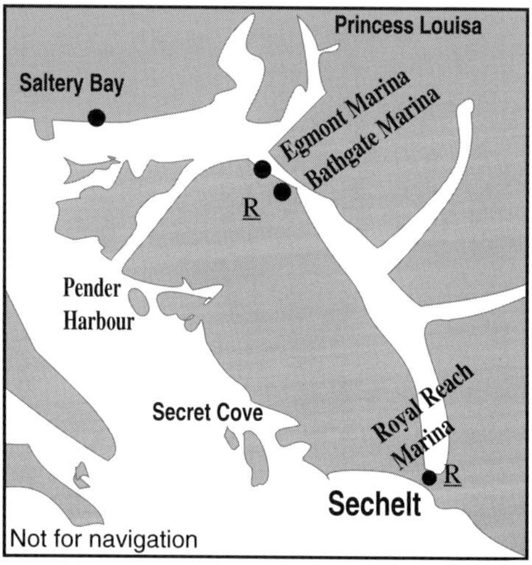

Marina services

Moorage.

Fuel: gas diesel,oils, two stroke oil, live/frozen bait at fuel dock.
Power 15, 30 amp.
Potable water. Ice, bait, tackle, fishing licences, book exchange, groceries.
Laundry, showers, washrooms. public phone ashore
rental boats available.

Entertainment.

Backeddy Marine Pub, summer long-weekend live entertainment.
Pool table. Nearby trails to Skookumchuck rapids.

Adjacent facilities:

Restaurant, store, scuba air station.
Dive and fishing charters.
Launch ramp. Campground–22 sites with power. View cabins. Hotel.
Kayak lessons and rentals.
Mountain bike rentals.

Sechelt Inlet

Royal Reach Marina and Motel

Charlie Shu
5758 Wharf Rd.,
Porpoise Bay BC. V0N 3A0
Phone: (604) 885-7844 fax (604)885-5969
Chart 3311, 3512

Marina services

Moorage.
Power 15 amp. **Water.**
Ice at marina and motel. **Laundry, showers,** washrooms at motel and ashore.

Entertainmnet.
Sechelt nearby has numerous centres, shops and facilities.
Walk to Sechelt village.

Adjacent facilities:
Pub, restaurant nearby.
Regular/scheduled flights.

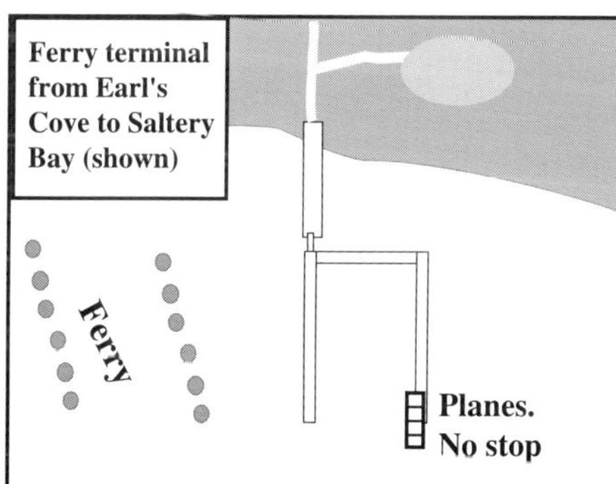

Saltery Bay,

Sunshine Coast
Fisheries & Oceans dock
Phone 487-9541
Chart 3514
Manager • Float length 133 m
Garbage • Lights •
Adjacent ferry dock–Egmont, Sunshine Coast.

Westview–Powell River

Beach Gardens Marina

Joan Barszczewski
7074 Westminster Ave,
Powell River, BC. V8A 1C5
Phone: (604) 485-7734 Resort 485-6267
Fax (604) 485-2343 Ph: 1 800 663-7070
Chart 3311, 3513 VHF 68

Marina Entrance.

At breakwater just south of Grief Point.

Marina services:

Fuel: Gas, diese., oil. ice.
Mechanic and services available from local and nearby marine operators. Repairs can be arranged.
Moorage: Large permanent marina with overnight moorage slips. Reserve.
Water at dock. Multiple outlets.
Power at docks: 30, 15 amp.
Laundry, showers, washrooms.

Customer services:

Hotel with full facilities. Sauna, pool.
Cold wine and beer store.
Grocery store, post office nearby.
Church/es in town. Fine dining restaurant in the hotel. Pub open 7 days a week.
Walking trails or road access. Some beachfront walks.
Fishing charters.
Scuba diving arrangements and charters–ask hotel for details.
Public pay phones ashore.

Entertainment.

Canoe Room marine pub.

Adjacent and nearby facilities:

Tennis, golf, hiking trails.
Shuttle service.
Shopping during season.

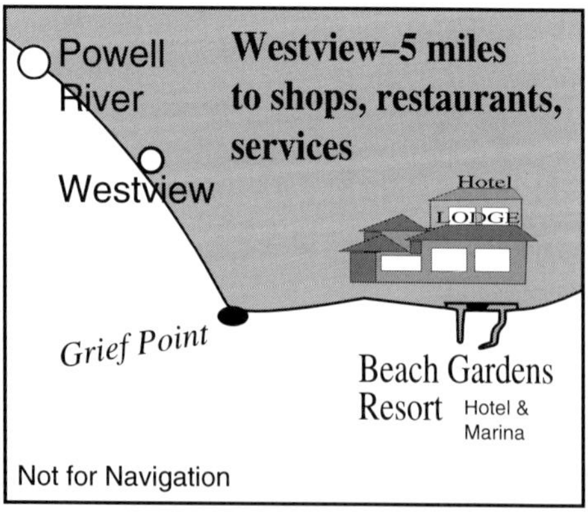

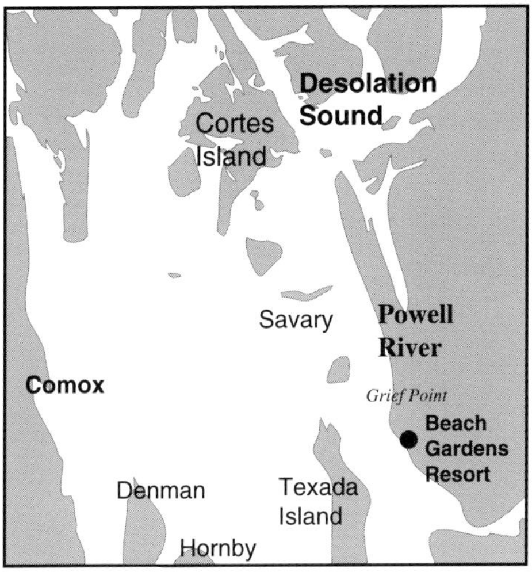

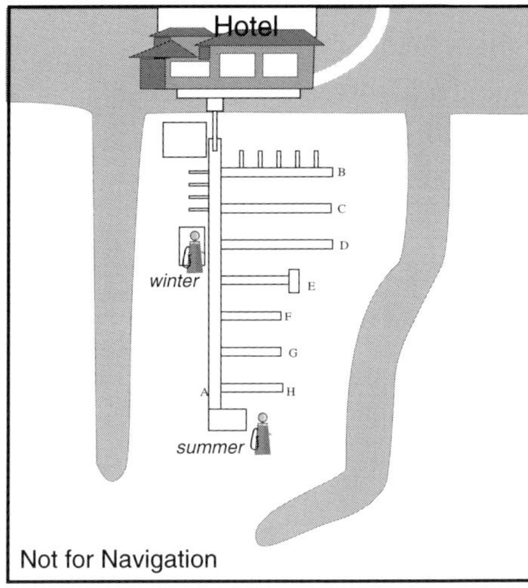

Beach Gardens Resort is one of the most popular stops on the Sunshine Coast.

The hotel/resort complex became a favoured stopover for cruising mariners in the mid 1970s when a breakwater was built in front of the hotel. The breakwater was later extended to accommodate resident boat owners in need of moorage. It became also a major stop for mariners en route to Desolation Sound. Monitoring the weather conditions at Grief Point has been made easy by the inclusion some years ago of Grief Point as a weather reporting station. The existence of Beach Gardens Resort at Grief Point provided the staffing required to observe and report up-to-the-minute conditions.

Each year scores of boats stop over in search of recreation at the spa, the fine cuisine for which the hotel restaurant has become famous and the tranquil setting of the facility. Fishermen and other vacationers fly into the area to spend time at Beach Gardens Resort or out fishing with one of the regular fishing charter operators based at the marina.

If you are travelling north to Desolation Sound or beyond, or returning from a trip in that area this marina is a worthwhile stop. It is a choice stop heading southbound especially if the weather is deteriorating. Rough seas down Malaspina Strait can cause much discomfort, particularly off Cape Cockburn. Restaurant facilities along with hotel spa and pool provide the type of comfort a family deserves after extended periods in the confines of a boat.

In summer time the fuel dock is usually located at the entrance to the marina and in winter it is moved closer towards the shore for protection from the swells that wash around the end of the breakwater. Some year-round marina tenants means that not all dock space is available for transient moorage and it is best to check with the dock manager before tying up. In boating season a full time staff working the docks will be awaiting your arrival. It is also best to check in by phone or call on VHF prior to your arrival. Reservations may be made well in advance or at the last moment depending on available space.

The next stop for fuel and moorage after Grief Point is Westview a few miles north en route to Powell River, Lund and Desolation Sound. Fuel is available south at Pender Harbour.

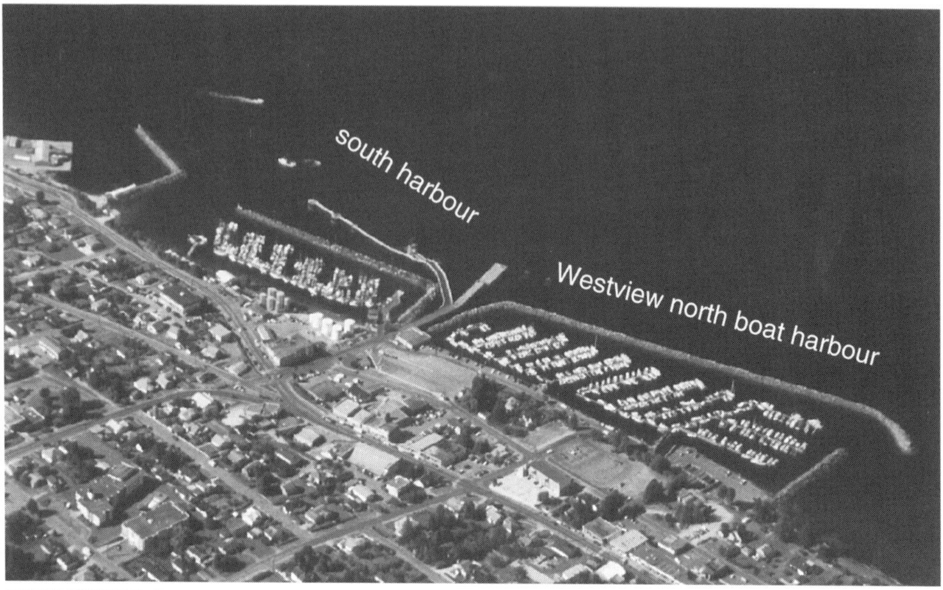

Powell River public docks

Westview (north) Boat Harbour

Transport Canada dock/Boat Harbour
Ph (604) 485-5244 (**no transient moorage**)
Chart 3311, 3536, 3513
Breakwater and marina
Launch ramp • Water • Lights • Power •
Near ferry landing, city restaurants, shops.
Fuel dock nearby: Gas, diesel, outboard mix, naptha, water, ice (in South harbour).

Powell River South Harbour

Transient moorage
Fisheries & Oceans dock
Chart 3311, 3563, 3513
Manager–Jim Parsons **(604) 485-5244**
• Float length 625 m
Breakwater • Grid • Garbage • Waste oil disposal • Water • Lights • Power • Public phone • Washrooms •

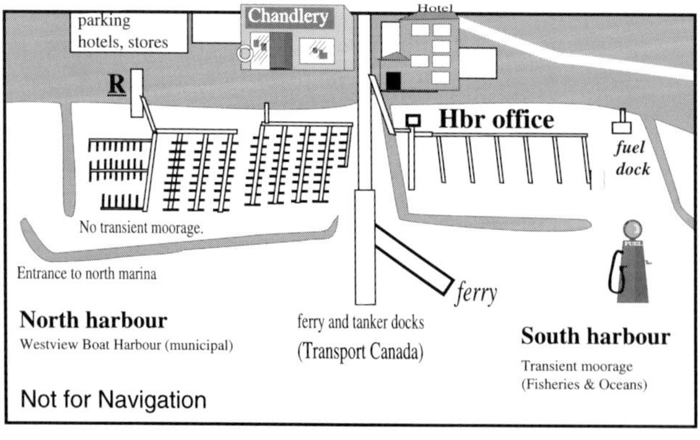

Fuel dock phone: (604) 485-2867

Near city restaurants, shops, hotels. Enter southern marina via fuel dock.
Fuel dock. Located inside breakwater. Gas, diesel. Outboard mix. Ice.
All services and facilities. fishing tackle, marine supplies nearby. Most located at north harbour.
Also up town.
Ferries adjacent to Comox, Vancouver Island and Texada Island.

Blubber Bay

Texada Island
Transport Canada dock
Chart 3311
Float length 13 m
Limited facilities and moorage. Best for overnight at Westview on Sunshine Coast shore opposite. Facilities also at Powell River.
Marina at Beach Garden Resort.

Lasqueti Island Hotel & Resort (False Bay)

Bruce Bird
Lasqueti Island, BC,
V0R 2J0
Phone: (250) 333-8846 Fax (250) 333-8897
Chart 3512, 3513, 3536

Marina services:

Fuel: Gas, diesel. Stove oil. Ice at local store. **Moorage** (limited) at nearby government dock. **Water** at dock.

Customer services:

Hotel with facilities including jacuzzi, gym, convenience store. Showers. Public phone. Washrooms. Restaurant–licensed.
Walking road access. Fishing and boat charters. Good scuba diving in nearby locations. Public pay phone.

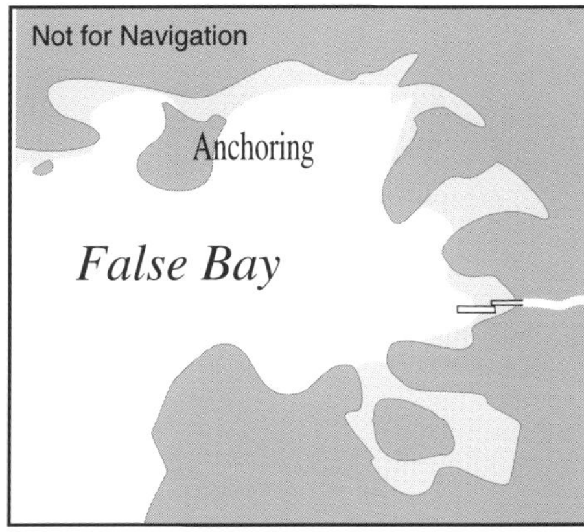

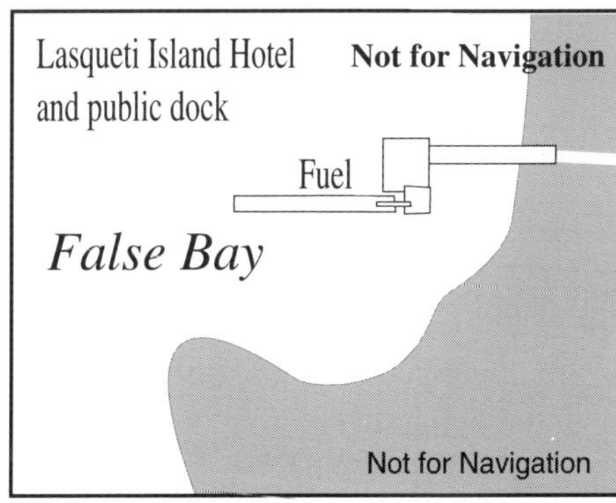

False Bay

Lasqueti Island
Transport Canada
Chart 3536, 3512, 3513
Float length 36 m • Aircraft Float • Garbage • Lights • Power •

Squitty Bay

Lasqueti Island
Fisheries & Oceans dock
Chart 3512
Float length 47 m • Garbage •
Walking, trails, island roads.

Entertainment.

Bike rentals. Fishing charters available.

Adjacent facilities:

Anchoring at end of bay–see local chart.

Lund

Above: The marina has upgraded facilities including an improved causeway and ramp adjacent to the marina entrance.

Lund Public Marina

(Fisheries & Oceans) Rosemarie O'Neil
Phone: (604) 483-4711
Chart 3311, 3538, 3513 VHF call 73

Marina services:

Moorage – over 500 feet for transient use. Launch ramp. **Water.** Lights. **Power**–20 amp. Public phone.
Adjacent Lund Hotel, restaurants, shops.
New launch ramp and access upgrade.

Lund Hotel (Fuel dock)

(604) 483-3199.
Fuel: Gas, diesel, **showers, laundry,** ice, propane. Paid garbage drop. Restaurant, pub–open 7 days a week.
General store, groceries, tackle, marine supplies. Charts. Some hardware.
Liquor agency. Nearby church/es.
Walking trails or road access.
Some beachfront walks. Boardwalk.
Fishing charters. Dive shop, air, rentals. Scuba diving arrangements and charters–see dive/kayak shop adjacent to hotel. Kayak rentals, water taxi service. Post Office.

Adjacent facilities:

Stores, coffee house, arts and gifts.
Restaurants. Ice cream, pizza. Bakery. Fresh, live seafood. Fish and Chips and breakfast kiosk. Farm market Sundays 10:30 to 2 pm.

Hoeger Yachting Ltd.

Box 73, 1444 Lund Highway, BC
V0N 2G0
Phone: (604) 483-9002 Fax: (604) 483-4914
email: *hoeger@prcn.org*
Moorage. Marine supplies, hardware, clothing. Repairs–Mechanic, electrical, wood, fibreglass. Emergency towing. Ways.
Pressure washing. Charters.

Composite photograph above shows a view of the Lund waterfront looking south. The hotel on the left overlooks the wharf and fuel dock. The public docks lie beyond the immediate foreground.

Public docks

Okeover Inlet (Arm)
Fisheries & Oceans dock
Phone (604) 483-2218
Chart 3559, 3538, 3513
Manager • Float length 35 m
Breakwater • Lights • Power •
Public phone ashore •
Near restaurant.

Finn Bay, Lund
Fisheries & Oceans dock
Phone (604) 483-4478
Chart 3311, 3513, 3538
Manager • Float length 49 m

Savary Island
Transport Canada dock
Chart 3311, 3538, 3513
Float length 11 m

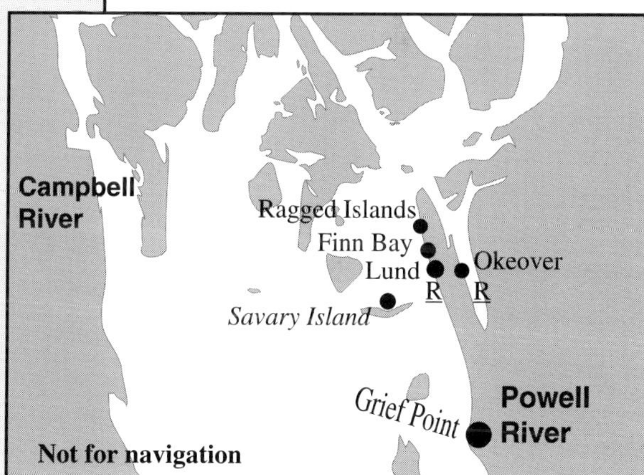

The Lund Hotel is a landmark on the Sunshine Coast where it sits at the northern end of Highway 99, an extension of the US I-5 freeway.

Lund is an interesting place, existing as the last town on the Sunshine Coast road. It has an historic hotel and numerous facilities including antique store, watersport and dive shop, grocery store selling a fair selection of marine hardware, a fresh seafood shop and an adjacent marine service centre and chandlery catering year round to the commercial and fishing trade. It's also a busy fuel stop before entering Desolation Sound. However there is a lesser known fuel stop at the Copeland Islands just a few miles north. Ragged Islands Marine has a pricing structure that makes you pleased to have stopped there for fuel.

Copeland Islands

Ragged Islands Marine

Wendy and Derek Cox
P.O. Box 22
Lund BC. V0N 2G0
Phone: (604) 483-8184
Chart 3311, 3513, 3538

Marina services:

Moorage: Overnight limited space.
Fuel: Gas, diesel, oil.
Water, snacks. Gifts.
Fishing tackle, licenses.

Customer Services:

Accumulate fuel points for lower prices.

Adjacent facilities:

Ragged Islands Marine Park. Anchoring. This is the entrance to Desolation Sound.

Ragged Islands Marine docks can accommodate a couple of boats overnight. The store has some snacks and supplies. It is a good fuel stop.

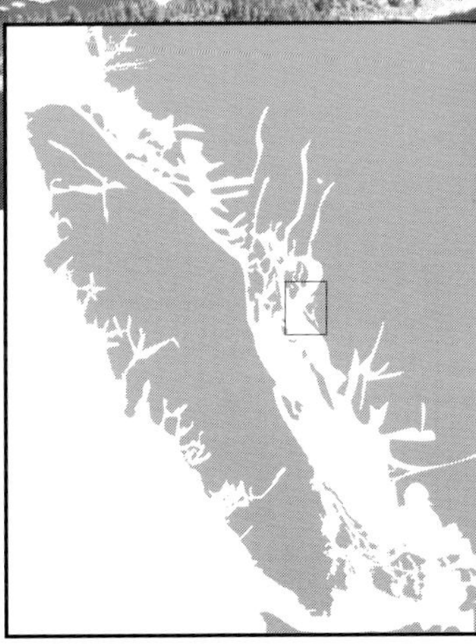

The Copeland Island (top) provide a protected waterway from Lund to the entrance of Desolation Sound. Ragged Islands Marine is off to starboard near the exit of Thulin Passage. Not far into Desolation Sound is Refuge Cove (above), a centre of activity for fuel and supplies.

Vancouver Island

Central east coast

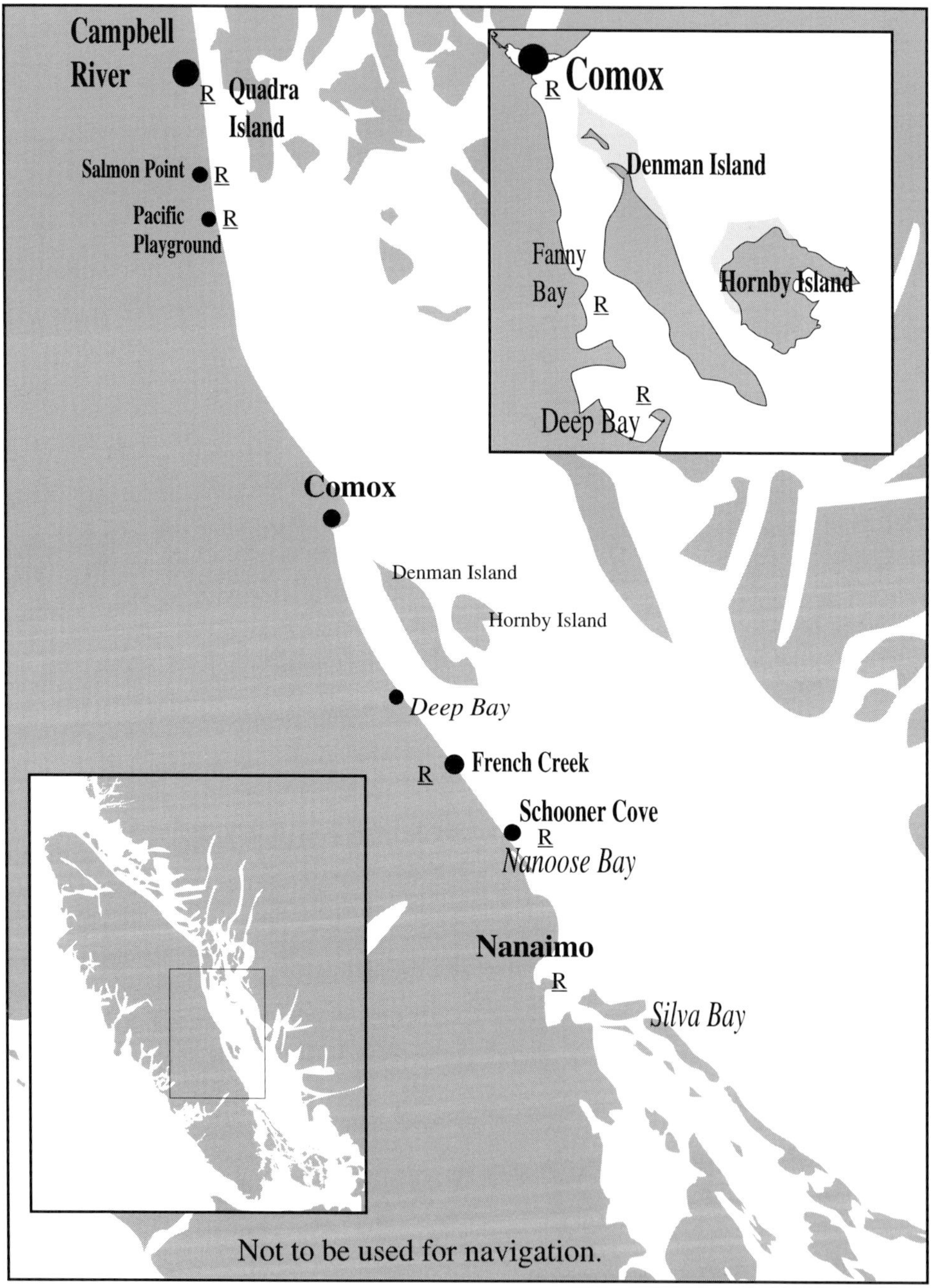

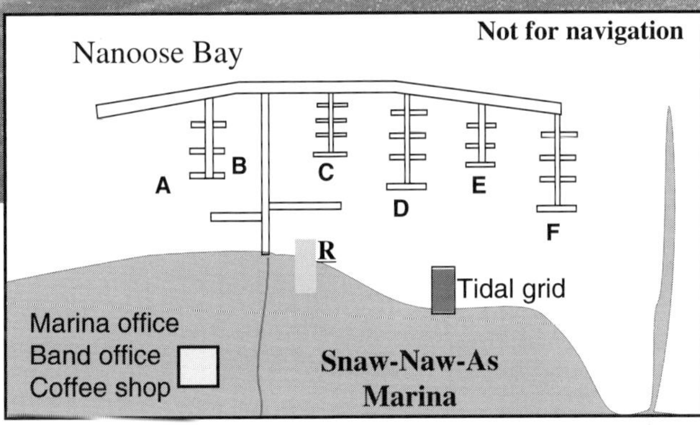

Above and Left: Snaw-Naw-As Marina, a local aboriginal band facility in Nanoose Bay, welcomes all transient boats.

Nanoose Bay area

Snaw-Naw-As Marina

(Located at entrance to Nanoose Bay)
209 Mallard Street,
Lantzville, B.C. V0R 2H0
Phone/Fax: (250) 390-2616
Fax (250) 390-3365
Chart 3459, 3512

Marina services:

Fuel: gas, diesel.
Water. Power. Showers.
Tidal grid, launch ramp, coffee shop. Telephone ashore. Garbage disposal. Fish cleaning. Porta Potti toilets ashore. Camping.

From Nanaimo go along the coast to Campbell River with a stop possible at Nanoose Bay or at Schooner Cove, one of the largest and finest facilities anywhere with hotel and all amenities including golfing nearby, good fishing and good scuba diving. Farther north up Vancouver Island's east coast is Comox with a vast set of docks, private and public. Comox Marina or Blackfin next door, with lots of transient moorage and restaurants ashore, a park and nearby stores and other facilities along with events such as Nautical Days each August will keep you entertained and enthused about the stop-over. En route to Campbell River you will find sheltered facilities at Deep Bay or French Creek.

Schooner Cove

Schooner Cove Resort

Hotel & Marina

Murray Hamilton
Box 12, Schooner House,
3521 Dolphin Drive,
Nanoose Bay, BC, V0R 2R0
Phone: Marina (250)468-5364
Hotel (250) 468-7691 Fax: (250) 468-5744
Marina reservations 1 800 252-5404
email: *info@fairwinds.bc.ca*
Web: *fairwinds.bc.ca*

Hazard: Drying rock in entrance to marina. Pass marker to port and keep close to floats.

Marina services:

Fuel: Gas, oils, service available–can be arranged at dock.
400 permanent and transient berths.

Chart 3512, 3459, VHF call 16 & 73

Water at dock. All berths.
Power at docks: 30, 50, 15 amp. All slips.

Customer services:

Hotel, accommodation, restaurant, lounge, pub. Fine dining. Coffee shop. Full fitness centre and spa.
Laundry, Showers, Washrooms.
Outdoor pool, hot tub. Public phones.
Marine supplies, fishing gear, licences, charts, bait, **ice,** books, gifts, snacks.
Cold beer and wine store. Bakery.
Nearby church/es: multi-denominational.
Post Office:
Courtesy shuttle to post office, banks, groceries. Scuba diving arrangements and charters–ask marina for details.

Part of the marina at Schooner Cove.

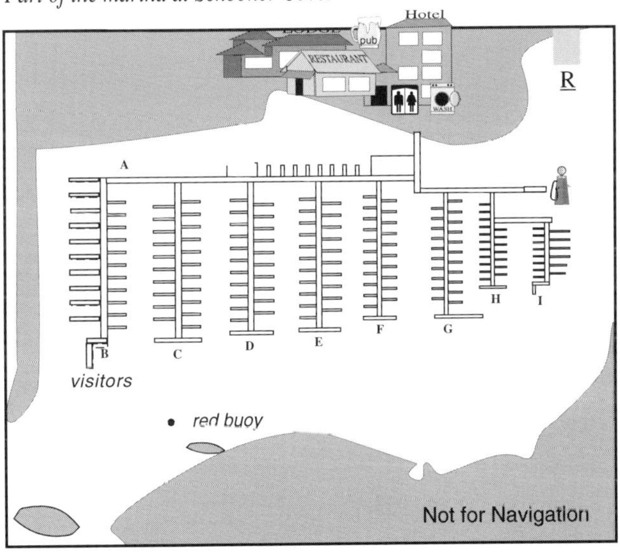

Schooner Cove is in one of the most beautiful settings on the coast. Nearby golf courses and developments have turned the area into a prized piece of real estate and the adjacent waters into prime fishing and scuba diving destinations. The hotel marina office (below) is adjacent to restaurants and other facilities, including marine supplies.

Walking: Road access walking or cycling. Nearby parks.

Entertainment.

Fishing excellent near marina.

Rental bicycles. Horseback riding, nature walks.

Adjacent facilities:

Courtesy shuttle to golf at Fairwinds 18 hole course, bare boat charters and fishing charters. Group facilities. **Launch ramp.** Picnic area overlooking marina.

Schooner Cove to Comox

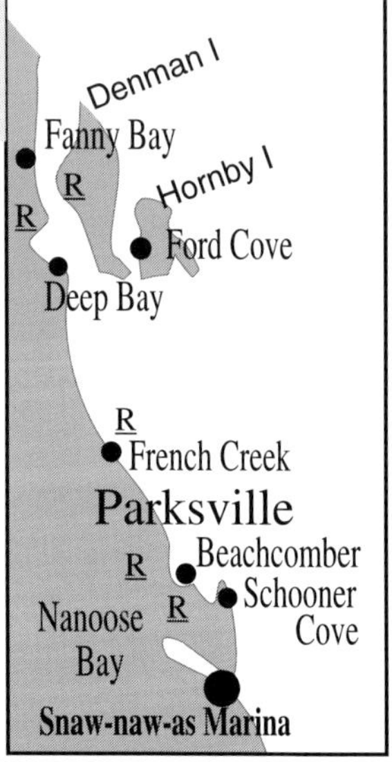

Beachcomber Marina

(Located in Northwest Bay)
Bill Loewen
RR #1 Box 21, Beachcomber
Nanoose Bay, BC, V0N 3A0
Phone/Fax: (250) 468-7222
Chart 3459, 3512. *Refer charts for reefs–enter between red and black buoys.*

Marina services:

Fuel: gas, diesel, ice (in summer).
Permanent and limited transient docks.
Water at dock.
Power at docks: 30 amp.
Launch ramp.
Anchorage across bay behind log booms.

Deep Bay,

Large, protected marina near Hornby and Denman Islands.
Chart 3513
Launch ramp.
Trailerable–**fuel**.

Ford Cove, Hornby Is

Photograph—opposite page

Fisheries & Oceans dock
Phone (250) 335-2169
Chart 3527, 3513
Manager • Float length 86 m
Breakwater • Grid • Garbage • Waste oil disposal • Water • Lights • Power • Public phone • Washrooms •

French Ceek

Please turn to next page

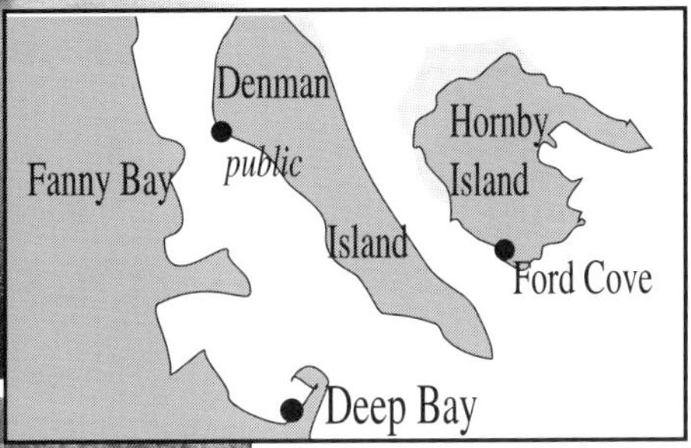

Left: Deep Bay. Use caution in adjacent shallow waters. Launch ramp access with store and fuel for trailerable boats. Public dock is near the ramp.
Below Ford Cove on Hornby Island.

Denman Island

Fisheries & Oceans dock
Chart 3527, 3513
Float length 24 m
Launch ramp • Breakwater •

Fanny Bay,

Fisheries & Oceans dock
Phone (250) 335-2513 Chart 3513
Manager • Float length 42 m
Breakwater • Grid • Power •

Chart 3512

Photograph above: French Creek. Bottom: A view of the French Creek docks.

French Creek

11055 Lee Rd
Parksville, BC,
V9P 2E1
Phone: (250) 248-5051
FAX (250) 248-5123
(pub) Fisheries & Oceans dock
Phone (250) 248-5051
1 977-4725 toll free
Chart 3512, 3513
Manager • Float length 804 m • Ramp • Breakwater • Grid • Garbage • Waste oil disposal • Water • Lights • Power • dry storage • rentals • repairs • moorage • public phone • ice • gas • adjacent restaurants, shops. Seafood store– prawns, crabs. Foot ferry to Lasqueti. Coast Guard station. **Launch ramp.**

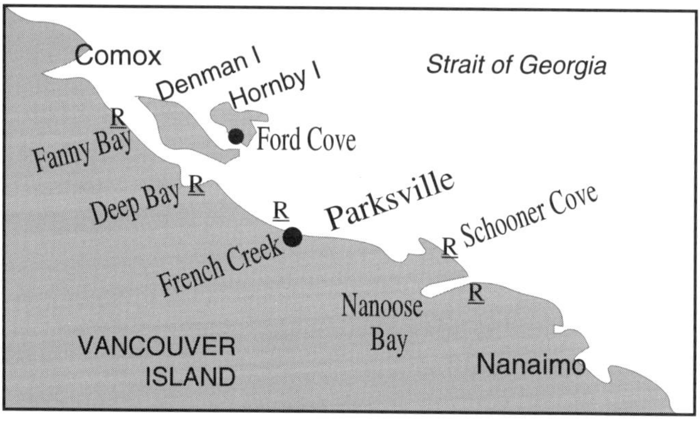

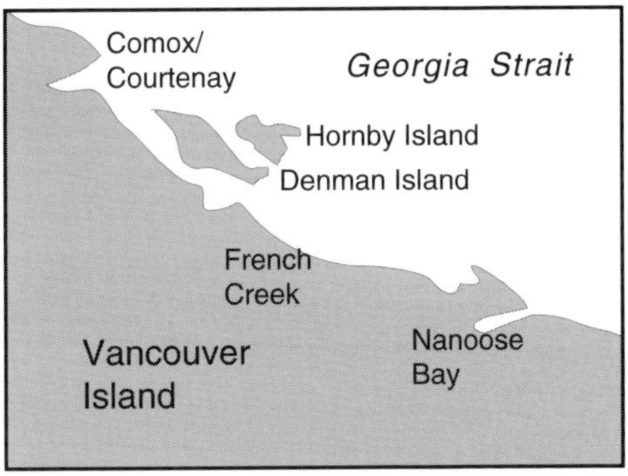

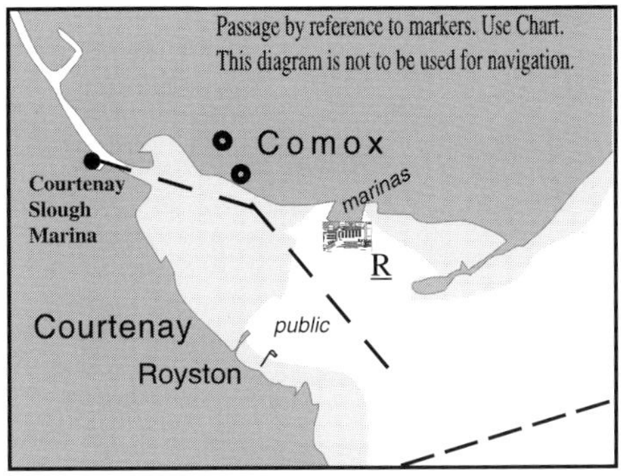

Comox–Courtenay

The Blackfin Pub & Marina

Sharlene Bentley
132 Port Augusta St,
Comox, BC, V9N 8G7
Phone: (250) 339-5030 (pub)
Phone: (250) 339-4664 (marina)
Fax: 339-4664
Charts 3527, 3513   VHF call 68

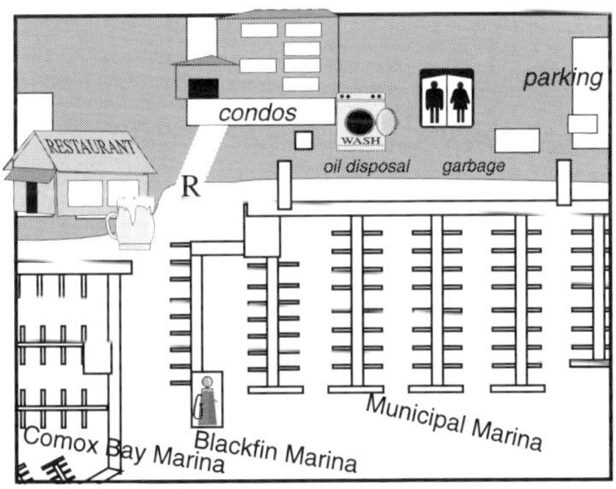

Marina services:

Fuel: Gas. Diesel. oils, outboard mix. Ice.
Moorage: Boats up to about 45 feet.
Water at dock.
Power at docks–15 amps.

Customer services:

Laundry, Showers, Washrooms.
Restaurant and pub.
Scuba diving arrangements and charters–ask hotel for details. Public pay phone.

Entertainment.
Town pier–pleasant for walking.
Nautical Days celebrated every August.
Picnic facilities at Marina Park.
Rental vehicles.

Adjacent facilities:

Boat charters. Golf, Shopping centre, liquor store, post office. Public dock. Garbage disposal.
Canadian Armed Forces base. HMCS Quadra camp at Goose Point.

Comox Bay Marina

1805 Beaufort Ave.,
Comox, BC, V9N 1R9
Phone: (250) 339-2930
Charts 3527, 3513

Marina services:

Moorage: Transient and permanent. Boats to all sizes.
Water at dock.
Power at docks: 50, 30, 15 amp.
Hydro-hoist available. Tidal grids.

Customer services:

Laundry, Showers, Washrooms. Ice.
Restaurant. Lunch, dinner. Plus breakfast at new motel.

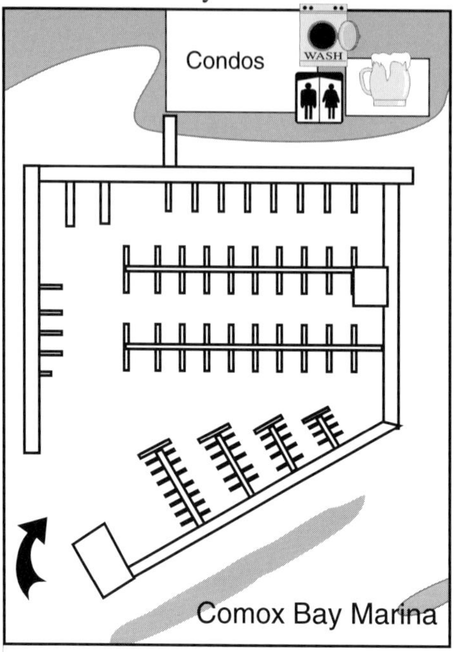

Comox Bay Marina (continued)

Walking: Road access, nearby park. Town of Comox. Public pay phone.

Entertainment.

Town pier–walk.
Nautical Days celebrated every August.
Picnic facilities at Marina Park.
Rental vehicles available.

Adjacent facilities:

Charters. Golf, shopping centre, liquor store, post office. Motel suites and condos.
Yacht club. Good view and location. Public dock. Garbage disposal.
Canadian Armed Forces base. HMCS Quadra camp at Goose Point.
Ferry to Powell River departs Comox.

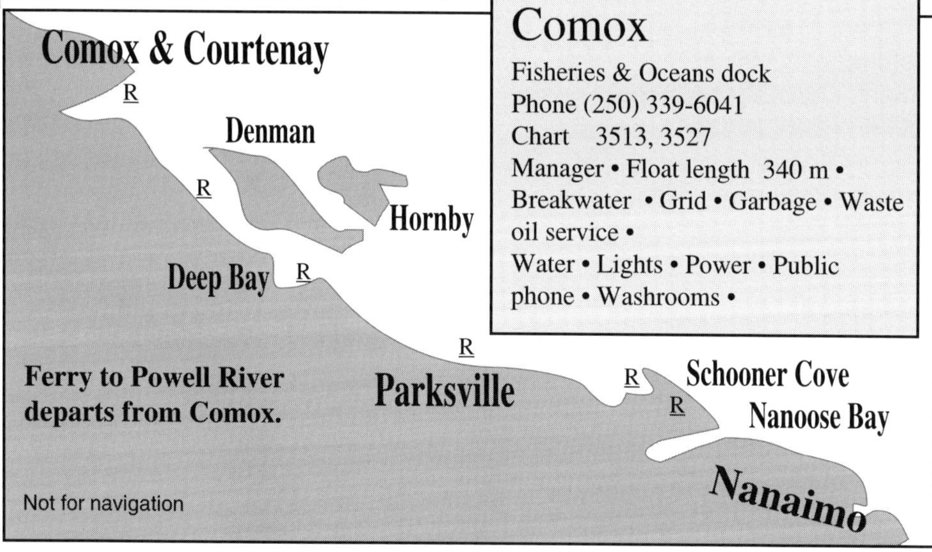

Comox

Fisheries & Oceans dock
Phone (250) 339-6041
Chart 3513, 3527
Manager • Float length 340 m • Breakwater • Grid • Garbage • Waste oil service •
Water • Lights • Power • Public phone • Washrooms •

Courtenay Slough

Fisheries & Oceans dock
Chart 3527, 3513
Manager • Float length 210 m
Garbage • Waste oil service •
Water • Lights • Power •

Comox Royston

Fisheries & Oceans dock
Chart 3527, 3513
Manager • Float length 140 m
Waste oil service •

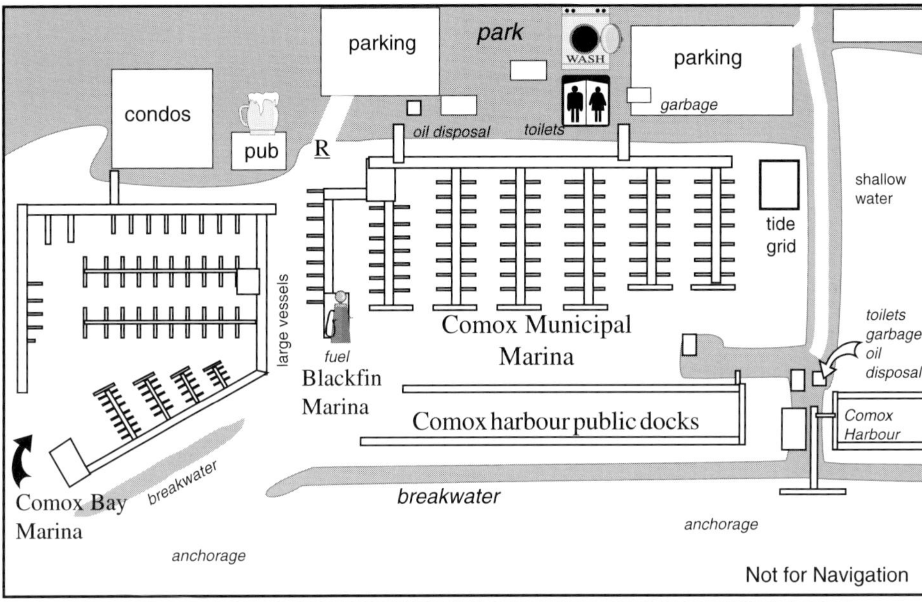

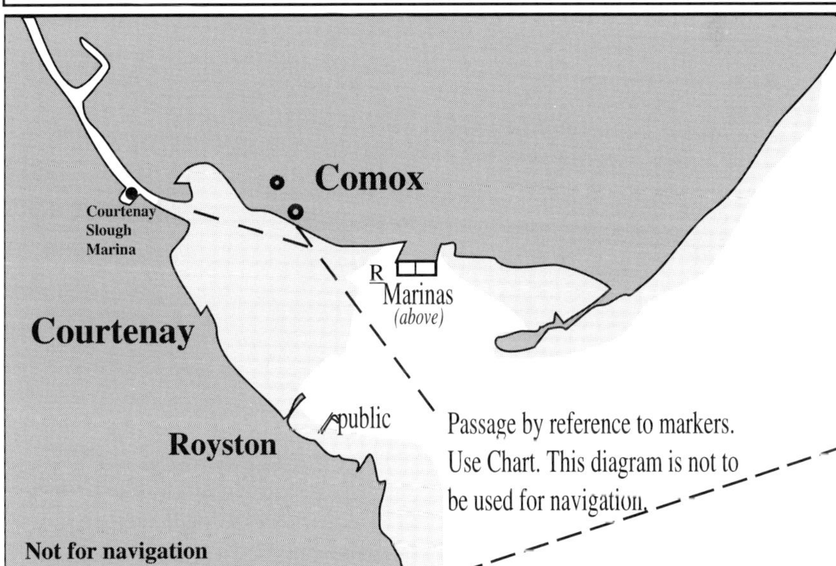

Comox Municipal Marina

Dave Hanson

Ph: (250) 897-2381. Town of Comox (250) 339-2202

Chart 3599, 3527.

Water, power, garbage disposal.

Moorage for small boats & a few large slips. Public phone. Used oil disposal. Tidal grid. Fishing pier and promenade. Adjacent anchorage. *Hazard: drying flats to north in harbour near pier.*

Courtenay Slough Marina

Limited moorage for small boats.

Phone: (250) 338-8124.

Water, power, public phone.

Hazard: Shallow river. Depths allow boat with shallow draft only–about 4 feet. Check chart 3599.

Walking: Uptown Courtenay.

Public pay phone.

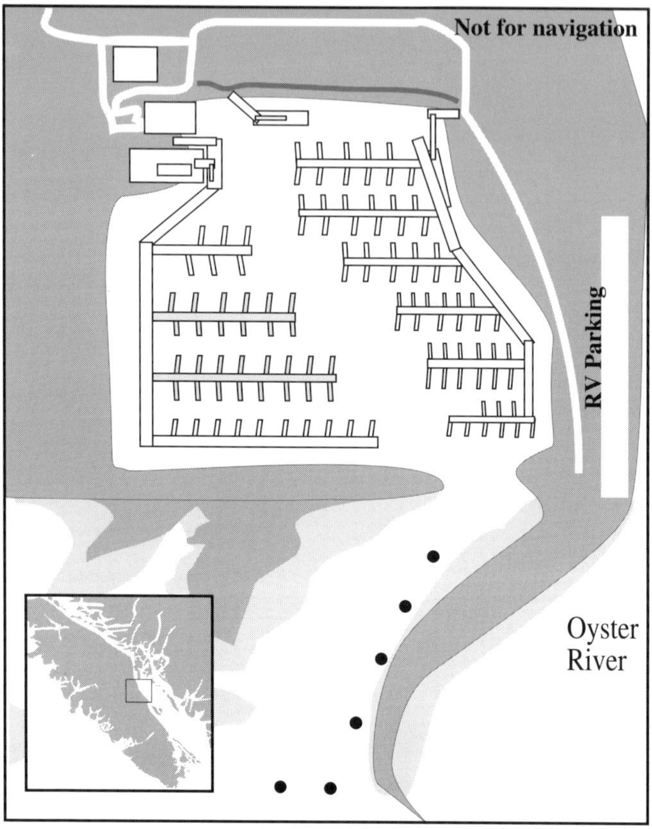

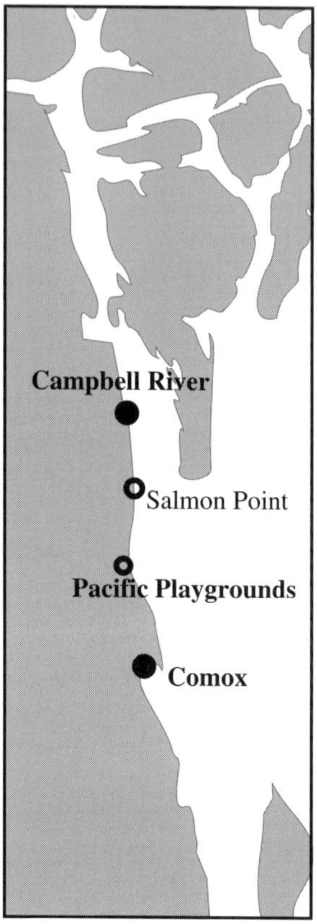

Comox to Campbell River

Pacific Playgrounds Resort

Walt Lengerke
9082 Clarkson Drive,
Black Creek, BC. V9J 1B3
Phone: (250) 337-5600.
Chart 3527, 3513 (Alongside Oyster River Mouth.)
Hazard–Channel may be entered at 4 foot (plus) tides. Follow pilings.

Marina:

Moorage. Sheltered basin. Numerous slips to 40 feet.
Fuel: Gas, diesel. **Water.**
Power–30 amp.

Services:

Marine store: Supplies, tackle, charts, fishing licenses, groceries. public phone. Resort facilities including heated pool, grassy play area. **Showers, laundry, washrooms.** Garbage disposal.

Entertainment and nearby facilities:

Golf, mini-golf, driving range, tennis, hiking roads and trails. Stores and restaurants. Scenery, sunsets and eagles. Nearby beach walks, bird watching. Fishing: guides available.

Diagram: Pacific Playgrounds Resort with its open water entrance. Its approaches are shallow but are clearly marked.

Cape Mudge

North to Campbell River

This coastal section is not well known to mariners from Vancouver and mainland points south. Most boats cruising to Desolation Sound and beyond travel out of Washington State and the Lower Mainland of British Columbia via the Sunshine Coast. Mariners from Victoria and Sidney tend to favour the Vancouver Island side and therefore know the coastline better.

After leaving Nanaimo, one can stop at Schooner Cove for a few nights of rest and recreation, knowing that golf is just a short courtesy bus ride away, or that there are some scenic roads and trails to walk. The salmon fishing out of Schooner Cove is exconvenient and the scuba diving surprisingly good, with such locations nearby as Mistaken Island and Sangster Island. There are resorts along the way that cater emphatically to sports fishing and these include Pacific Playgrounds on the Oyster River and Salmon Point Resort approaching Campbell River.

Typically these resorts cater to RV campers and lodge guests, but they also welcome overnight mooring guests who have full acces to their many resort facilities. After even one day of boating it is relaxing to climb into a spa and unwind, better still after several days. Or to stretch one's legs on shoreside nature trails, on the tennis courts or the golf courses available. Guides are available for fishing and this is a good idea if you really want to catch salmon.

As you travel up this coast you have the option of stopping at places such as Deep Bay, Nanoose, Fanny Bay, French Creek and across at Hornby or Denman Islands. These islands are beautiful and the waters that surround them worthy of slow cruising by just for observation and sightseeing. Flora Islet attracts boatloads of scuba divers in search of the famed six-gill sharks that dwell in the near shallows. These creatures are rarely seen in BC. They are encountered frequently by divers at Hornby Island although seldom if ever by others. Dally around the islands while water conditions are calm but look for protected overnight moorage at Vancouver Island shoreline marinas when the wind is up. Ford Cove on Hornby is the only hope of shelter in an adverse blow. Dock space is limited and anchoring out may not provide a comfortable night's sleep. If you have a shore boat visit the lighthouse on Chrome Island

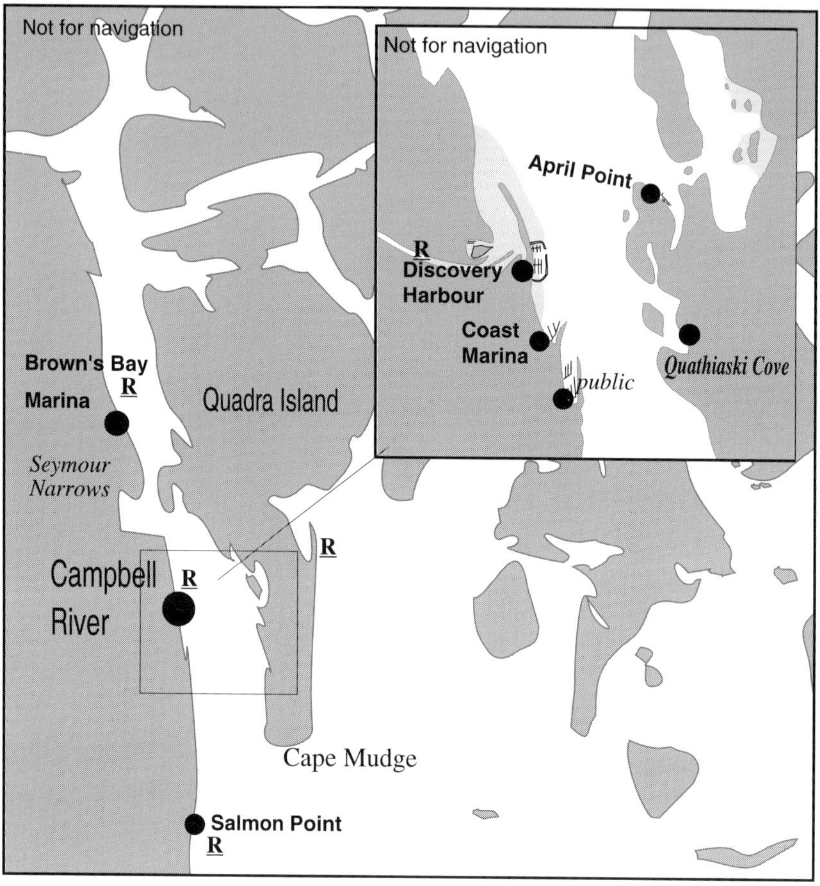

and look for the petroglyphs on a rock face just below the lighthouse. Many artists live on Hornby Island and if you can reach the island centre you are bound to find some excellent works of art by local people.

Entering Campbell River you pass Cape Mudge on your starboard side but be mindful of the tides and winds as a huge sea can occur off Quadra Island when conditions are stormy or just windy at maximum current velocities. The currents in the passage can be strong and treacherous too so take care to keep out of the most current swept areas if you are running in a slow boat. There are numerous mooring options in Campbell River, from the government docks at Quathiaski Cove or in Campbell River with their limited or typical public facilities to a night aboard tied up to the docks of April Point Fishing Lodge on Quadra or Brown's Bay Marina just beyond Seymour Narrows. There are large marinas such as Discovery Harbour or the Coast Marina alongside the ferry dock for Quadra. This latter facility is operated by the Discovery Inn and the large marina located there offers many facilities as well as fuel, charter fishing and scuba excursions. Campbell River is renown for its outstanding scuba diving and several operators are located there to cater to the needs of visiting divers.

Anchoring in Gowlland Harbour offers overnight protection from most wind and current conditions.

Campbell River

Salmon Point Resort

2176 Salmon Point Road,
Campbell River, BC. V9H 1E5
Phone: (250) 923-6605

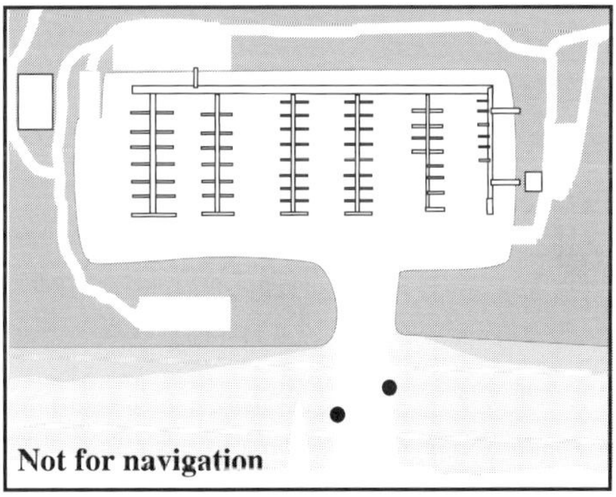

Chart 3527, 5313

Hazard: Narrow channel into marina. Shallow. Proceed between floats off breakwater. Call–ask marina for entrance guidance.

Marina:

Sheltered moorage for up to 200 small boats (to 26 ft.). Limited transient moorage.
Fuel: Gas, limited water, power. Phone.

Services:

Resort with many amenities–for resort, moorage and RV guests. Restaurant.
Hot tub, heated pool.

Nature trails, tenting. Laundromat. Restrooms, showers.

Fishing guides and charters. Garbage disposal. Fish freezing and packaging.

Entertainment and nearby facilities:

Golf, mini-golf, driving range, tennis, hiking roads and trails. Stores. Scenery, sunsets and eagles. Nearby beach walks, bird watching. Fishing.

After putting into Campbell River for supplies and perhaps a break from solitude, many visiting boats head for the marina tucked into Quadra Island on the other side of Discovery Passage. It is located beyond the resort at April Point.

April Point

Fishing Lodge Resort & Marina

Dock manager
P.O. Box 1, Campbell River, BC. V9W 4Z9
Phone: (250) 285-3621 (Marina)
285-2222 (lodge) Fax (604) 285-2016
Charts 3312, 3540, 3539, 3513 VHF 73

Marina services:

Ice, live bait. Mechanic and services on site. Fibreglass repairs. Outboard motors sales/ service. Launch ramp.
Moorage. Large permanent marina with overnight or extended moorage available. Reservations recommended in mid summer. Marine shop (250) 285-3177
Water at dock. Multiple outlets.
Power at docks: 20, 30, 50 amp. Multiple outlets.
Laundry, showers, washrooms.

Customer services:

Lodge with full numerous amenities.
Fishing guides and charters.
Restaurant. Breakfast, lunch, dinner. Open 7 days a week. Coffee shop. gift shop.
Conference rooms.
Nearby church/es.

Walking trails or road access. Some beachfront walks. Horseback riding.
Scuba diving arrangements and charters–ask lodge for details.
Public pay phones ashore.
Scooter rentals. Taxi service.
Accommodations at lodge and bungalows.
Transport to golf courses. Airport limo.

The April Point Lodge with its fishing lodge floats out front. Restaurant service at the lodge. Inset: Farther in the cove is the 'Yacht Club' moorage for overnight guests.

Approaching marina, pass red buoy to your right. Consult your chart.

April Point 'Yacht Club' docks

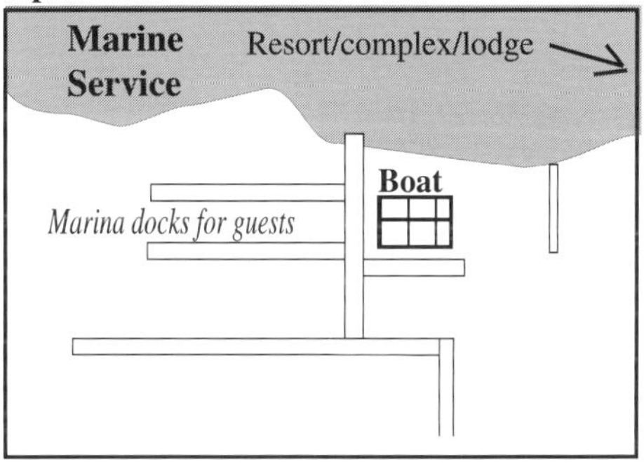

Entertainment.

Barbecue on Saturday nights.
Cable television.

Adjacent facilities:

Kenmore Air regular flights.
Liquor, grocery, arts and crafts and other stores including post office nearby.
Hiking trails at Rebecca Spit Provincial Park on Quadra Island.
Also walk island roads, trails and beaches.
Indian Museum and Cultural Centre.
Philips Marine. Fax (250) 285-2407

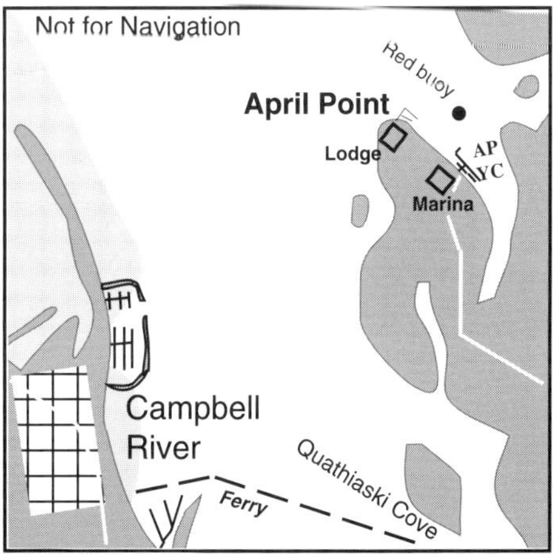

Photograph: Osvaldo Brasca

1. The Coast Marina

formerly Sportfish Centre

Osvaldo Brasca

975 Shopper Row, Campbell River, BC. V9W 2C5. Phone: (250) 287-7155
Fax (250) 287-2213
Chart 3540, 3539 VHF call 73

Marina services:

Campbell River Chevron Marine Station (Dave Facey)
Campbell River,
BC. V9W 6Y4
Phone: (250) 287-3319
Fax: (250) 287-8767
Chart 3540, 3539 VHF 73

Fuel: Gas, diesel, stove oil, outboard mix, **Moorage.** 2700 feet docks. **Power:** 15, 20, 30, 50 amp/220 volt. **Water.** Garbage disposal.
Dockside marine services.

Customer services:

Laundry, showers, ice, bait.
Public phone ashore.
Marine charts, books, fishing gear, licences, boat supplies, CNG, kerosene and stove fuels. Light refreshment (snacks).
Nearby church/es: multi-denominational.
Walking: Foreshore Park with walkway to scenic downtown Campbell River.

Entertainment:

Scuba diving arrangements and charters–ask marina for details.
Fishing excellent in general area .
Fishing charters, boat rentals.

Adjacent facilities

Ferry adjacent. Restaurants and all services–located in centre of town.

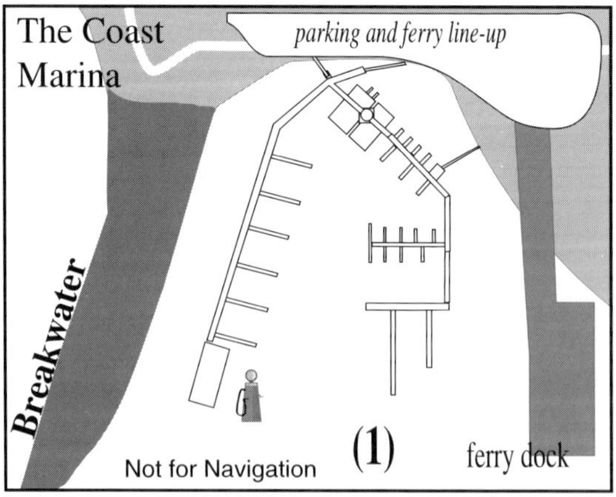

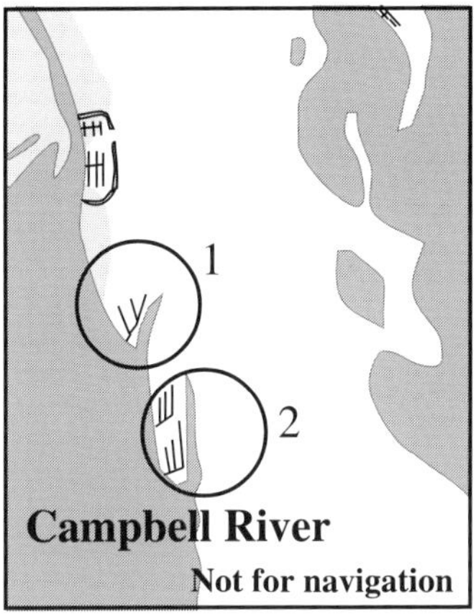

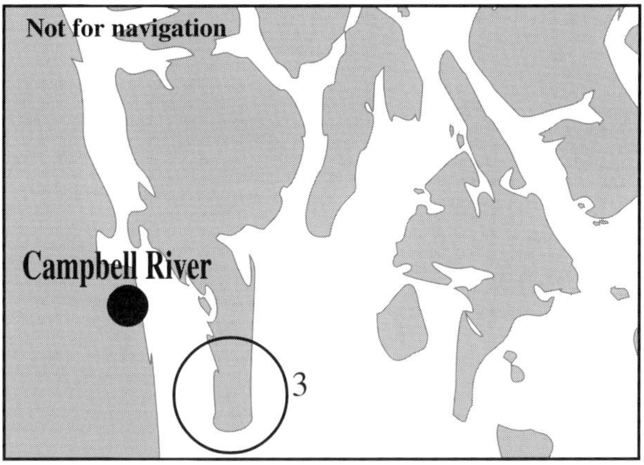

2. Campbell River

Fisheries & Oceans

Phone 287-7931
Chart 3540, 3539
Manager • Float length 1033 m
Breakwater • Grid • Garbage •
Waste oil service • Water • Lights •
Power •
Public phone • Washrooms •
Showers •
Fuel • Gas, diesel, CNG, stove oil.
Ice, bait, tackle, charts.
Campbell River town adjacent.
Also ferry to Quadra Island.

3. Cape Mudge

Fisheries & Oceans dock
Phone 285-3622
Chart 3540, 3539
Manager • Float length 60 m • Breakwater • Marine Ways • *The float is to the left of Cape Mudge–shown on a calm day in the photograph below.*

Campbell River is a busy place. There are resorts at this international sportfishing playground that cater solely to fishermen, offering guided salmon fishing and tournaments. Some offer no transient moorage while others have an open door to overnight boating stops. The velocity of water surging through Seymour Narrows just north of the town makes for interesting navigation, even dangerous so be mindful of tides and current, especially when venturing through the narrows or around the bottom of Quadra Island and its infamous Cape Mudge during gusty winds and swift moving waters.

Fuel up at Campbell River or across the way at Quathiaski Cove, or just beyond Seymour Narrows at Brown's Bay, because you may not find fuel too conveniently for a while if you are going north up Johnstone Strait, unless you turn off to Blind Channel or head into Desolation Sound. Blind Channel not only has fuel but also an excellent restaurant, general store and other facilities including a liquor agency. Fuel is available also at Heriot Bay on the east side of Quadra Island as well as at Gorge Harbour, Manson's Landing and Refuge Cove.

Dock A is for 18 foot boats, B is for 20, C & D for 24, E & F for 30, H is for 36, I and J are for 40 and K for 100 footers.

Discovery Harbour Marina

1400 Weiwakum Road,
Campbell River BC. V9W 5W8
Phone: (250) 287-2614
Fax: (250) 287-8939
Chart 3540, 3539
VHF call 73
ask for Discovery Harbour Marina

Marina services:

Fuel: Esso dock. Gas, diesel.
Moorage: 150 berths for transient moorage. Boats to 100 feet and over.
Power at docks: 15, 20, 30, 50, 100 amp.
Water.
Dockside marine service can be arranged.

Customer services:

Laundry, showers, ice, bait.
Public phone ashore.
Nearby churches: multi-denominational in Campbell River.
Walking: Road access walking or cycling. Vehicle rentals in town.

Entertainment:

Scuba diving excellent in area. Strong tidal currents, use local dive operators as guides. Fishing excellent in general area.

Quathiaski Cove,

Quadra Island.
Fisheries & Oceans dock
Phone (250) 624-2244
Fuel dock: (250) 285-3212
Chart 3540, 3539
Manager • Float length 195 m
Launch ramp • Garbage • Waste oil disposal •
Lights • Power • Public phone •
Adjacent–Shipyard, ***Fuel dock**: Diesel, gas, outboard mix, water, propane, oil. Shops nearby. Adjacent ferry to Campbell River.

***Coastal Mountain Fuels**
Fuel. Bait, snacks, fishing gear.

Quathiaski Cove

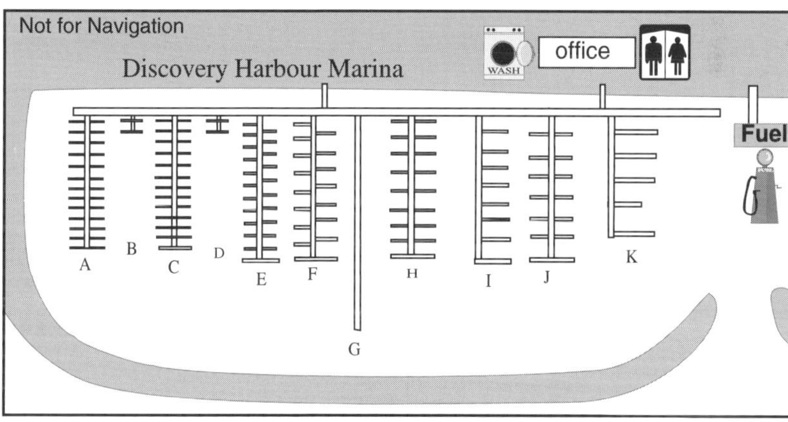

Discovery Harbour Marina is a large marina with a vast breakwater. It is on aboriginal native land and is operated by the local native Indian band. Opposite Campbell River is Quadra Island with its Quathiaski Cove serving as the ferry landing.

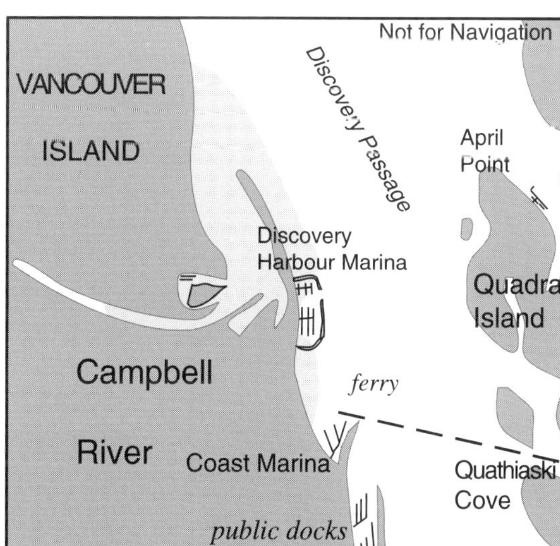

Brown's Bay Marina

Jim and Julie Camp
Box 668, Campbell River
BC, V9W 6J3
Phone: (250) 286-3135 Fax 286-0951
Chart 3539, 3513

Marina services:

Fuel: Gas, diesel, oils.
Propane. Oil change facility.
Moorage. 1800 feet transient dock. Boats to over 100 feet.
Power at docks: 30, 20, 15 amp.
Water. Ice. Garbage disposal.

Customer services:

Bed and breakfast accommodation.
Laundry, showers, ice, bait.
Public phone ashore.
Store. Marine charts, tackle, licences, foul weather gear. Fish cleaning. **Launch ramp.**
Floating restaurant/cafe.
Boat rentals. Fishing. Wildlife viewing. Guides available. Charters.
Walking: Road access walking or cycling.

Entertainment:

Fishing superior in general area near harbour. Outstanding scuba diving at Campbell River and vicinity.

Adjacent facilities:

Ripple Rock RV Park–many facilities.

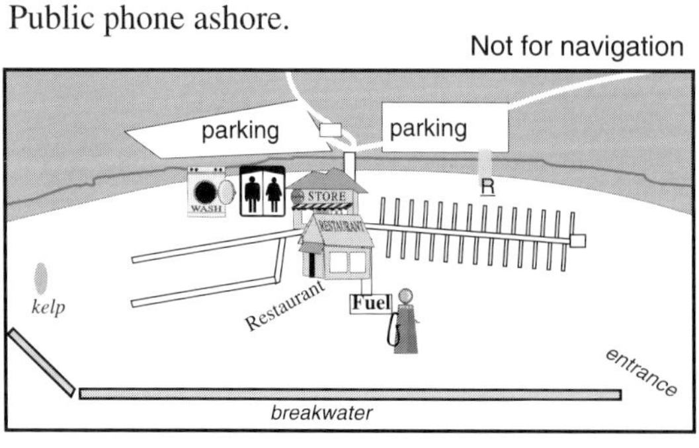

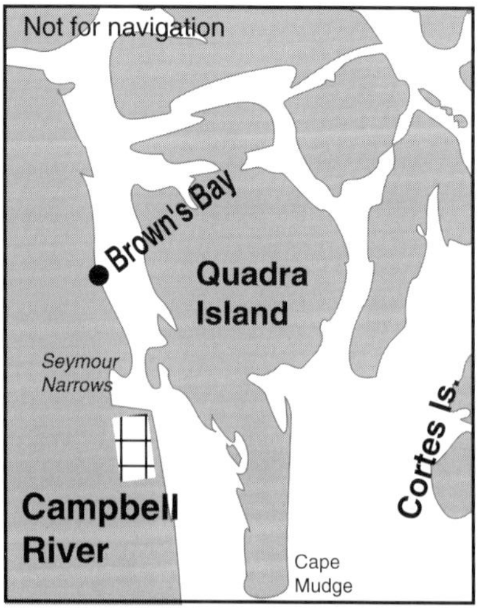

Desolation Sound

Central Coast Playground

Once in Desolation Sound you will not be stuck for fuel because there is a full service fuel stop at Refuge Cove. This facility is a centre for all boating needs to serve the cruising mariner. It has liquor, fresh produce, groceries, frozen foods, books, charts and more. The crafts shop, coffee bar and hamburger stand on the property above the marina make an ideal place to wander or sit in the sun and enjoy the ambience of being out boating.

Desolation Sound is a popular place to anchor for days on end in the summertime. From the many coves and bays of Grace Harbour or Prideaux Haven to Pendrell Sound and Walsh Cove, or Roscoe Bay, Theodosia Inlet, Van Donop Inlet and Squirrel Cove to name a few,

Refuge Cove Store, a busy stop in the summer months.

one can find the ideal place to set up home aboard for a few days or play musical moorages and move from one to the next as one spends a summer vacation in this warm water oasis in BC. Move early in the day to avoid difficulty in finding a place to drop anchor in some of the busier bays.

To the west is Heriot Bay for moorage and fuel and Drew Harbour with its anchorage behind Rebecca Spit. Kayaking is drawing increasing number to this area each year. Go to Surge Narrows from Heriot Bay, or cross over and spend the rest of your vacation at Gorge Harbour, one of the most sheltered large bays in the area, with a fine marina and restaurant to keep you in comfort for your stay. There is fuel at Gorge Harbour, a store, all facilities plus petroglyphs on the sheer rock face at the entrance. In the vicinity you may want to stop in for a stroll at nearby Whaletown. Dock space is limited. *See **Anchorages and Marine Parks** for more information on Desolation Sound.*

Gorge Harbour

Gorge Harbour Marina & Resort

Glen and Verlie Carleton
P.O. Box 89 Whaletown,
BC. V0P 1Z0
Phone: (250) 935-6433 fax 935-6402
Chart 3538   VHF 73 CB 18

Marina services:

Fuel: Gas, diesel, oil.
Moorage: Guest moorage.
Water at dock. Multiple outlets.
Power at docks: 30, 15 amp.
Multiple outlets.

Customer services:

Grocery store. Fishing licences, bait, tackle. Ice, books, gifts, charts, propane, postage stamps. **Laundry, showers, washrooms.** Public pay phones ashore. Boat rentals. Private rooms with showers and baths. Good scuba diving nearby. Charters. Fish cleaning station.

Entertainment.

Adjoining campground with good facilities. Trail to roadway. Walk to rustic Whaletown. Art Gallery *en route.*
Video rentals. Restaurant dining.
Adjacent facilities:
Licenced restaurant on property–breakfast, lunch and dinner (May 1–September 30). Arts and crafts and museum on island. Campground. RVs. Scooter, car, bicycle, canoe and kayak rentals.

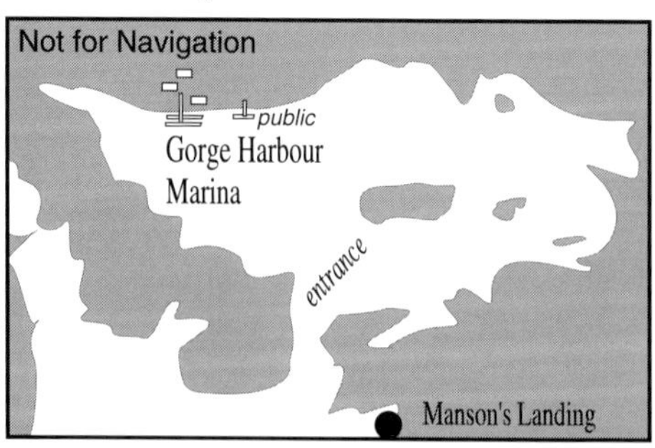

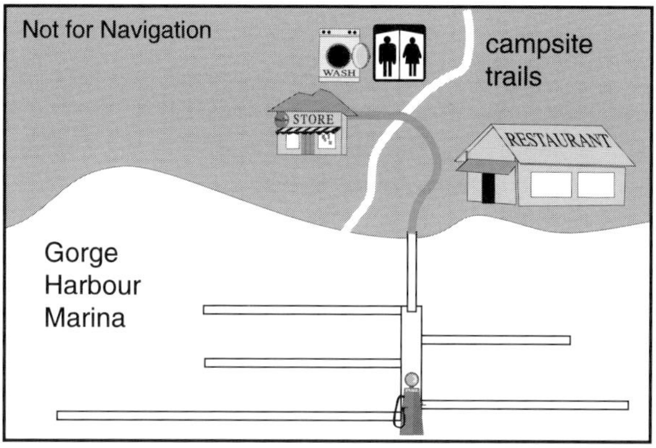

Below: The entrance to Gorge Harbour is narrow and subject to current. There are petroglyphs on the rock face to port. Opposite: Gorge Harbour Marina. Lower left and right: Manson's Landing and Whaletown.

1. Manson's Landing

Cortes Island. (store at Squirrel Cove)
(250) 935-6361
Irvine and Doreen Redel
Joanne Prestley
Chart 3311, 3538 VHF 73

Hazard: Exposed to westerly winds. Shallows on north approaches.

Marina services:
Limited moorage at government dock. See *Anchorages and Marine Parks.*

Alternative services:
Groceries at Squirel Cove or nearby at Gorge Harbour. Road to store and cafe.

Entertainment.
Walking on island roads. Beaches. Lagoon.

Adjacent facilities:
Marine Park.

2. Whaletown

Cortes Island, BC V0P 1Z0
(250) 935-6562 fax 935-6624
George Frost
Chart 3538

Hazard: Rock near government dock. See chart 3538.

Marina services: *No garbage.*
Transient, limited moorage at public dock. Dock space no charge (1998). Rafting likely.

Customer services:
Post office (Mon, Wed, Fri). Store: groceries, bakery goods, fresh produce, ice cream cones, milk, frozen foods, ice, water, tackle, bait, charts, novelties, gifts. Video rentals.

Entertainment.
Walking on island roads, swimming, scuba diving, fishing, kayaking, bird watching.

Adjacent facilities:
Ferry to/from Quadra. Public phone.

Heriot Bay

Heriot Bay Inn

Tom and Julia Pearson
P.O. Box 100, Heriot Bay,
BC V0P 1H0
Phone: (250) 285-3322 Fax 285-2708
Chart 3538/9 VHF 73
email: *heriot@oberon.ark.com*
http://oberon.ark.com/~heriot/index.html.

Marina services:

Fuel: Gas. Diesel. Oil. Propane.
Moorage, repairs, service available nearby.
Water at dock.
Power at docks: 30, 15 amp.
Fish cleaning station.
Boat rentals.

Customer services:

Restaurant–patio service (seasonal), pub (all year)–The Logger & The Fisherman. Kayak rentals. Fishing licences, bait, tackle. Store: Ice, books, gifts, charts.
Laundry, showers, washrooms.
Public pay phones ashore. Boat rentals.
Private rooms at Inn. Cottages. RV Park (cable TV). Hook-ups, tents.
Fishing, sight-seeing charters. Good scuba diving nearby. Bicycle rentals at marina.

Entertainment.

Historic Heriot Bay Inn and its classic pub. Walking–roadway and some beach access. Also at Nearby Rebecca Spit. Kayaking, adventure packages (ask for information at office).

Adjacent and nearby facilities:

Large grocery store, liquor store and mini shopping centre–delivery to boats. Arts and crafts.
Campground. RVs. Government dock.
Anchorage at Drew Harbour inside Rebecca Spit at Rebecca Spit Marine Park.

Heriot Bay

Quadra Island
Fisheries & Oceans dock
Phone (250) 285-3958 Chart 3538, 3539
Manager • Float length 204 m
Launch ramp • Garbage • Lights • Power • Public phone • Near marinainn, ferry to Cortes Island, park, shops.

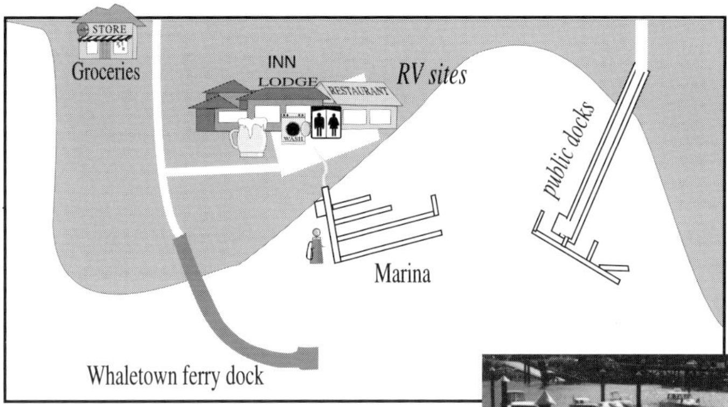

Heriot Bay Inn caters to boating. It has lodging, restaurant, marine bar, a spacious marina, store and fuel.

Not for navigation

Rich in history and charm, this cruising destination offers convenient provisions, ideal fishing conditions, and a compelling atmosphere.

If you are heading north during your summer cruise and stop en route at Heriot Bay, chances are you will remain right there for the rest of your holiday.

If you are fond of beachcombing, or strolling along country lanes, nearby Rebecca Spit and the immediate environs of Heriot Bay provide ample opportunity. If lounging on a sunny veranda appeals to you then you may get no farther than the sun porch at the Heriot Bay Inn. Indeed, this charming hostel evokes an ambience of a long-gone era. One writer described the Inn of Edwardian Times:

> *The beer parlour still flourished, offering relaxation to weary loggers and fishermen, but a gentler clientele was served by the remainder of the elegant rooms. There were 19 bedrooms and an upstairs dance hall, as well as a wide veranda on the second floor, used for summer dances. In this era –circa 1912!, hanging flower baskets decorated the lower veranda, which served as a tea-room, and young girls clad in white uniforms waited on the guests. There was an aviary of rare canaries, and oldtimers say a pet seal was kept in a pool behind the hotel.*
>
> *(Doris Anderson)*

The Inn today is still the centre of attention. For most visitors to Heriot Bay it is the focal point of the history of Quadra Island. The present building–several hotels over the years have occupied the same location–has a pub and restaurant, so yachtsmen can splice the main brace or dine out a short stroll up from their boats. The Inn provides float space, and offers fuel and basic provisions, mainly catering to the needs of fishermen who come from all parts of this continent, many religiously every year. The Inn also maintains a popular and spacious recreation vehicle campsite. The history of Heriot Bay is detailed in Bill Wolferstan's cruising guide, **Desolation Sound,** and in Doris Anderson's **Evergreen Islands.**

The Heriot Bay Marina is one of the few stops in the area where you can dispose of garbage. The fuel dock carries gasoline, diesel and engine oils as well as fresh water. Up at the Inn there are laundry and shower facilities. The 500 feet of mooring space is easy to access and is not subject to heavy current action. A marine facility at Quathiaski Cove, on the west side of Quadra, is equipped to do servicing and repairs.

Using Heriot Bay as a base, one can either spend time at the marina or at the adjacent government dock or at anchor in the lee of Rebecca Spit in Drew Harbour. Joyous summer experiences can include fishing and swimming or sunset strolls along the seaside paths of the A-class park that extends the length of the Spit. From your tender, you can troll or mooch the east side of the spit or several other nearby hot spots. A short walk up the main ferry road you will find a large and well stocked supermarket with adjacent coffee shop and craft stores.

Heriot Bay is located at the gateway to the Octopus Islands. A short run through Surge Narrows takes you to that increasingly popular summer cruising destination with fascinating inlets, bays and coves to explore en route. You are also close to Cortes Island and a Whaletown visit or a run down through Uganda Passage, which divides Cortes from Marina Island.

The nearby entrance to Gorge Harbour is a magnificent gateway of

Top and centre: Cortes Bay public dock, where wharfinger Bill brown gives regular visitors a warm welcome.
Left: Aerial view of Cortes Bay shows public dock to left, entrance to right.

The public dock in Cortes Bay. No private marinas are open to public use. A Seattle yacht club owns the major marina in the bay. It is for the exclusive use of members.

sheer rock on one side and wooded incline on the opposite shore. Indian petroglyphs can be found on the rock face of the north side. There is a restaurant with a fine reputation located inside Gorge Harbour. It is open and busy with boat crews and locals during the summer season. A small supply store on the marina property carries a variety of groceries and supplies. Another store is located at Surge Narrows. It carries a wide variety of goods including, groceries, snacks and coffee, hardware and some marine supplies.

Fishing and scuba diving in the general area are popular pastimes. Fishermen who visit the area frequently know the location where red snappers abound, and the size of some of the catches indicates a bonanza. The salmon caught on the Drew Harbour side of Quadra Island are not quite as impressive as those landed at Campbell River, but neither is the size of the tide rip.

One could easily leave one's boat tied up at Heriot Bay or at anchor in Drew Harbour and take the quick ferry ride across Sutil Channel to Cortes Island for a day's excursion. It is not recommended that you leave an anchored boat unattended, particularly where the anchorage is subject to large tidal changes and wind conditions.

The ferry lands at Whaletown and the road system on Cortes is ideal for more quiet country strolls.

When travelling from Lund to Heriot Bay via the west side of Marina Island refer to Canadian Hydrographic Service Chart 3565 or the Desolation Sound Chart Book number 3312 to avoid the shallows off Sutil Point and Marina Reef. They are marked by buoys, but be cautious. Running up through Uganda Passage on the east side of Marina Island is a pleasant alternative and don't fear the narrow waterway. It is well marked, but again follow your chart carefully and maintain a course to the correct side of the buoys and channel markers.

Anchorages and marinas

Facilities in Desolation Sound are scattered through and around the islands that comprise this lovely, popular cruising destination. Most replenishment stops are located west of Refuge Cove while Refuge is the main centre of supply to Desolation Sound. From Stuart Island in the north one could travel straight down Lewis Channel and stop in at Refuge Cove, or go west and take the options of Heriot Bay, Gorge Harbour, Whaletown, Manson's Landing or Squirrel Cove. To the south from Stuart Island is Lund with its mainland community and facilities comprising hotel, store and all services. In the central and eastern sections of Desolation Sound are the numerous anchorages and coves that most tend to favour for their stay. West of Refuge there are even more beautiful spots to anchor and shelter from winds and weather.

Near Cortes Bay it may still be possible to tour Wolf's Bluff Castle, complete with dungeon, racks and torture cell.

Refuge Cove

Located in the heart of Desolation Sound, this facility has been catering to pleasure boaters as long as there has been pleasure boating on the coast. Owners Colin Robertson and Norm and Adrienne Dowler welcome you warmly to their 2000 odd feet of dock space. Overnight moorage is secure and available with electricity, water and services includings fuel, ice, propane, groceries, fresh produce, liquor and post office. Adjacent to the store a small kiosk offers hamburgers, chips and other light food preparations. A bake shop and gift shop alongside offer coffee and treats on a cosy sundeck. As a small island community the residents ask that no garbage be left at the dock.

Cortes Bay

The dock is wide and sturdy although not anchored at the deep end. It is controlled by a wharfinger and shared with local residents owning pleasure and commercial craft. Space is limited and many yachtsmen favour anchoring out in the bay. Cortes Bay is windy and often appears to be more wind-swept than the waters outside. In recent years boat owners walked up the road to tour Wolf's Bluff Castle, a fairytale structure with an old Europe rustic atmosphere complete with dungeon, built by the owner Karl Triller, on a secluded two-and-a-half acre homestead property. No pets or smoking allowed inside the castle. Check with the wharfinger, Bill Brown, whether tours are still available. You will find him in his house at the public dock. Ask about the craft stores nearby. There are two yacht club out-stations in the bay which do not permit non-member moorage.

Squirrel Cove

The government dock allows for brief stops. Ashore the store offers groceries, fresh produce, ice, automotive supplies, fishing gear, bait and licences. visit also the craft shop on the property. There are telephones at the head of the dock as well as garbage disposal bins.

Manson's Landing

This exposed stop has a government dock with limited space, usually fully occupied by local and fishing boats. It is now a designated marine park. The store at the head of the dock has been moved to a new location on the island. There are other stores on the island including the one at the public dock at Squirrel Cove.

Gorge Harbour

The facilities at Gorge Harbour provide all one needs for a prolonged stay at the marina. The shoreside restaurant is excellent and the marina owners, Glen and Verlie Carleton, are friendly and courteous. The marina offers a reliable fuel supply and other marina services to transient guests.

Whaletown

The store at Whaletown has ice, groceries, produce, fishing gear, licences and much more for passing mariners. The post office is located nearby. There is limited space at the dock, which is protected from weather almost always. A rock in the bay is well marked, as is the one at the entrance.

Heriot Bay

Owners Tom and Julie Pearson offer lodging at Heriot Bay Inn, dining in the lounge and on the terrace. The pub has a classic island atmosphere. The marina offers moorage, fuel, water and power and other amenities, while groceries and produce are at the supermarket a short walk up the road.

Nearby anchoring is available at Drew Harbour in the protection of Rebecca Spit. Some windy conditions occur and boats most sheltered will be those tucked into the curve at the end of the spit. A drying area occurs half way along the beach so mind the depth when anchoring.

Left: At the store on the wharf at Surge Narrows. Below: The dock at North Rendezvous Island and the lodge with a view towards Raza Island. It is located 25 miles south of Big Bay near Surge Narrows and Refuge Cove.

Surge Narrows

Doug and Teresa Beyerstein
P.O. Box 31, Surge Narrows
BC V0P 1W0
Phone: (250) 285-3643 Cell 285-6962
Charts 3312, 3537, 3539

Moorage limited. Small dock with aircraft float. Post Office on dock. Open Mon, Wed, Fri 1 to 4 pm.

General store: Island fresh produce, baked goods and provisions. Coffee. Snacks. Some hardware and marine supplies. Limited water available.

This is a public dock. Additional private docks adjoining store are planned.

Passage through Whiterock Pass to Rendezvous Islands is always possible.

Rendezvous Lodge

Roy and Darlene Bugeaud
P.O. Box 309, Quathiaski Cove
BC. V0P 1N0
Cell phone: (250) 287-0318 Fax 203-1123
Charts 3539, 3541 VHF 12
email: *rbugeaud@online.bc.ca*
Web: *www.anglingbc.com/rdv/home.html*

Moorage. Laundry, showers, toilets. 240' of useable dock, used mostly by transient boats with dinner reservations at lodge. Use of hot tub included. Meals include wine. Rooms available.

Entertainment: Hiking trails, coves, beaches, oysters. Kayaks available.

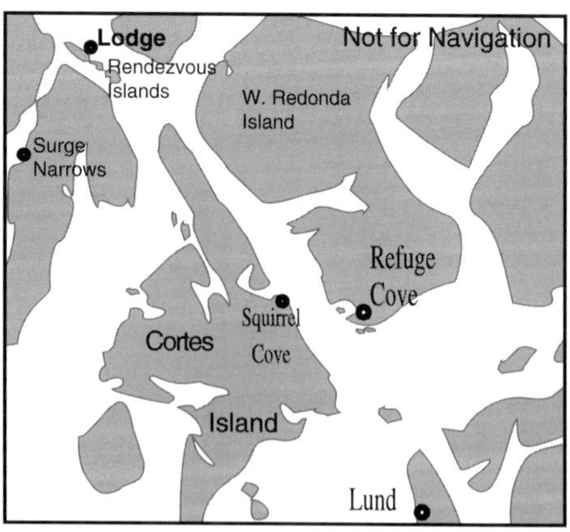

Dock at Squirrel Cove and inside the cove.

Squirrel Cove (Cortes Is)

Irvine and Doreen Redel
Phone (250) 935-6327
Chart 3555, 3538 VHF 73
Fisheries & Oceans public dock
Manager • Float length 60 m
Garbage • Lights • Power •

Marina services:
Temporary moorage at government dock.
Garbage drop.
Customer services:
Showers, laundry. Groceries. Public phone. Baked goods and fresh produce, milk, frozen foods, ice, water, tackle, bait, charts. Propane. Liquor. Gifts. Marine supplies.
Entertainment.
Hiking. Island roads. Beaches. Lagoon.
Adjacent/nearby facilities:
Squirrel Cove Marine Park. Anchorage. Arts and crafts shop.

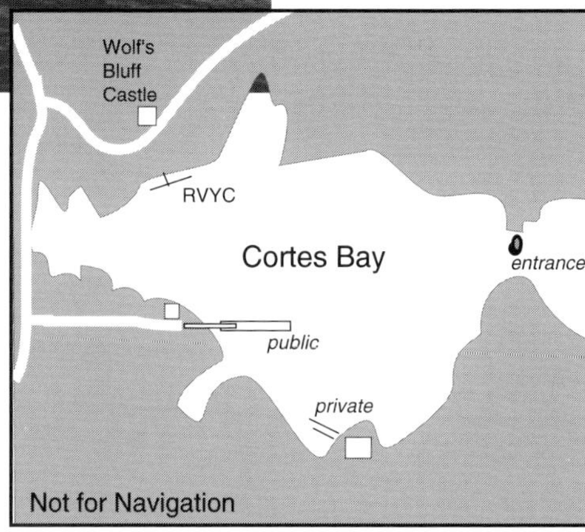

Cortes Government Dock

Cortes Island
Bill Brown (Wharfinger)
Chart 3555, 3538
Cortes Bay photopgraph left
Hazard: Entering bay–keep rock and day marker at entrance to starboard (passage south of marker).

Marina services:
Moorage at government dock.
About 200 feet of dock, wide and solid.
Windy at times in bay.
Power at dock: 15, 20 amp.
Garbage disposal.
Customer services:
Public phone. Arts and craft store on shore.
Entertainment.
Hiking. Island roads. Possible tour of Wolf's Bluff Castle–Information from wharfinger.
Adjacent facilities:
Yacht club station. Anchorage.

Refuge Cove

Chart 3555, 3538 VHF 73

Refuge Cove

Colin Robertson & Norm Dowler
Refuge Cove, BC V0P 1P0
Phone (250) 935-6659
crobert@oberon.ark.com
web: www.boattravel.com/refuge

Marina services:

Fuel: Gas. Diesel, oil, propane.
Moorage: Government dock and private. Overnight moorage. 2000 feet of docks.
Water at dock–multiple outlets.
Power at docks: 15 amp.

Customer services:

Laundry, showers, washrooms.
Propane, ice, books, gifts charts,
Grocery store, post office. Liquor.
Public pay phone on shore.

Adjacent facilities:

Gifts, arts and crafts, coffee bar, hamburger kiosk.

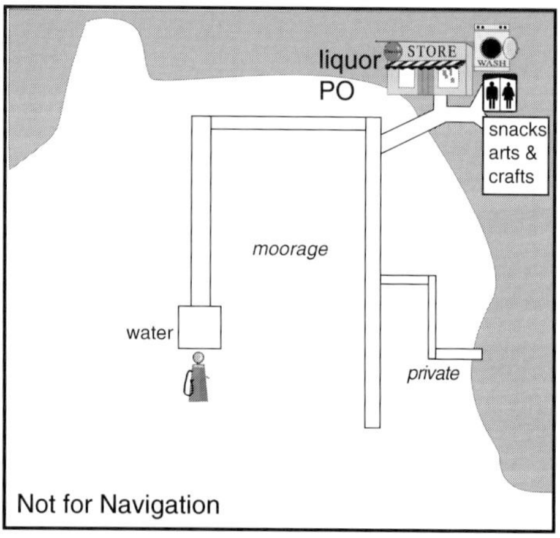

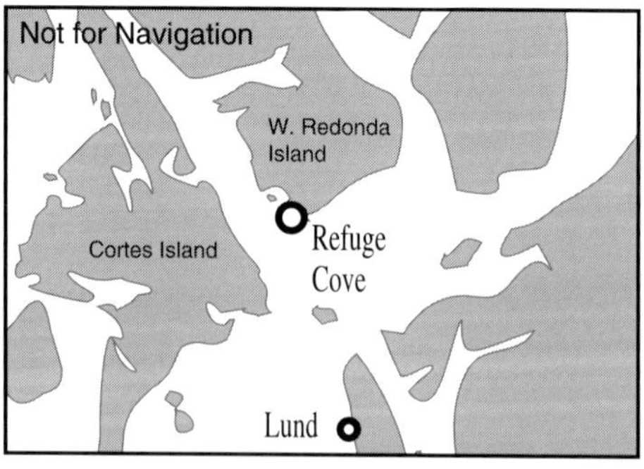

Above: looking onto the docks at Refuge Cove from the patio of the coffee bar next to the general store. L: Among the many anchorages in Desolation Sound, Grace Harbour is close to facilities and sheltered from wind and weather.

Oasis in Desolation Sound

En route to all parts of Desolation Sound, Refuge Cove is a busy stop during the brief summer season. It affords replenishment of everything from fuels (including propane) and fresh water to liquor, fresh produce, bread, meat and groceries. Charts, books and a limited selection of clothing are also sold at the store, as well as ice and fishing supplies. The store is built on a barge and positioned high and dry above the high-water mark. It and the fuel dock are run effectively to accommodate the heavy traffic of the short summer season and survive the balance of the year.

One of the burdens of summer is the threat of fire when forest conditions are very dry. Water shortages are sometimes threatened but seldom occur. The lake that drains into Refuge Cove is the supply of domestic water and the local population sometimes finds it difficult to prevent transient people from misusing it. The dock does, however, have adequate water for conservative use with outlets at various points for mooring customers.

It amazes me that Captain Vancouver wrote of the entrance to Desolation Sound as having a very inhospitable general appearance, especially considering he was there in the month of June. He may have encountered bad weather and was becoming depressed as suggested by Lester Peterson in **Raincoast Chronicles First Five.** Yachtsmen entering Desolation Sound from the direction of Lund via the Copeland Islands are usually awed by the beauty of the islands and waterways. They also often encounter what Captain Vancouver must have discovered—when the tidal streams from north and south meet around Kinghorn Island, turbulence can be quite strong.

If Captain Vancouver needed to encounter some semblance of the type of civilization from which he had come in order to restore his equanimity, then it is a pity Refuge Cove had not yet been established as a trading centre in Desolation Sound. If he had not been in need of provisions, then he could have accommodated himself with a shower or used the laundry facilities. Or he might have been able to browse through the souvenir and arts and crafts shop near the store. He would not, however, have been able to dispose of *Discovery's* garbage at Refuge Cove. To drop garbage he would have had to cross to Squirrel Cove on the opposite shore of Lewis Channel, where a dockside disposal bin is located. But Captain Vancouver could have saved himself much anxiety in his search for safe overnight anchorage by tying up *Discovery* along any section of Refuge Cove's 2000 feet of berthage, although the 530-ton *Discovery* would have left little room for the rest of us. Or he could have anchored in the northernmost bay of the Cove just near the Refuge Lagoon outlet.

To some yachtsmen, Walsh Cove may be just another ordinary anchorage at the perimeter of Desolation Sound. It is, however, a particularly pleasant and significant anchorage.

Located at the upper reaches of Waddington Channel—which separates East and West Redonda Islands—Walsh Cove is ideal for laying over at night prior to a run up Toba Inlet or a jaunt to Big Bay from Desolation Sound. It is a corner away from the main holiday traffic in the more central anchorages of summer's most popular sound.

It is not far from the supply store at Refuge Cove, and close to the warm summer waters of Pendrell Sound. Much of the traffic, therefore, arrives late in the day and leaves early in the morning, which means that arriving yachts should experience no trouble in finding a spot to anchor.

Walsh Cove, tucked into a steep rocky nook on the west side of Waddington Channel, is sheltered by the tiny Gorges Islets. Their protrusion above the water a short distance off the northern sweep of shoreline almost totally encloses the cove, protecting it from most winds (strong northerly winds can find their way in!), and the wash of yachts cruising by. An adequate channel allows only small vessels to slip in and out of the anchorage safely on the inside of the islets. The islets themselves are separated by a shallow, partially drying channel and a natural pool which attracts snorkellers and swimmers in summer. The water there reaches comfortable temperatures, although it's never quite as warm as nearby Pendrell Sound.

Desolation Sound
21 Anchorages:
1 Prideaux Haven
2 Laura Cove
3 Melanie Cove
4 Tenedos Bay
5 Pendrell Sound
6 Walsh Cove
7 Grace Harbour
8 Isabel Bay
9 Theodosia Inlet
10 Von Donop Inlet
11 Drew Harbour
12 Octopus Islands
13 Carrington Bay
14 Squirrel Cove
15 Evans Bay
16 Quartz Bay
17 Teakerne Arm
18 Gorge Harbour
19 Thurston Bay
20 Cortes Bay
21 Roscoe Bay
Check tide, current and wind conditions for safe overnight anchorage.

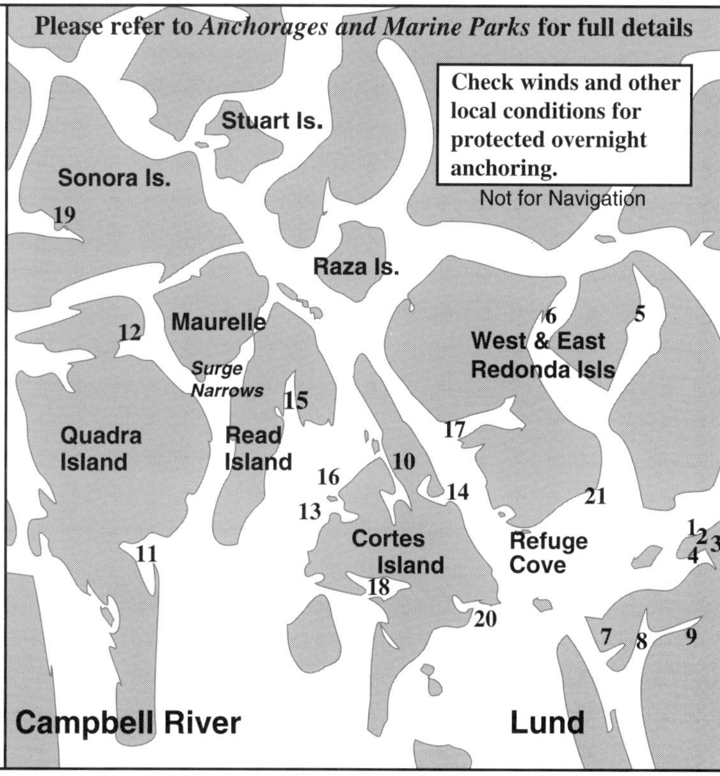

It is easy to become sidetracked in the vicinity of Walsh Cove or en route to a particular destination. Ovenight prawn traps set in about 40 fathoms on the east side of Waddington Channel opposite Allies Island may yield a return sufficient to enjoy a meal or at least a nice appetizer. A short distance up Toba Inlet's northwest shoreline there is a magnificent, high waterfall cascading down the rock face of the shore.

In 1792 the Spaniard Valdes, just leaving the inlet after exploring it to its head, encountered the British about to enter and advised them not to bother going up. "It goes nowhere," I imagine the Spaniard saying. It was definitely not the northwest passage that he knew was the main objective of the British expedition.

At Waddington's narrow entrance there are two sets of pictographs located at Butler Point opposite the north side of Gorges Islets.

Walsh Cove may appear to be incidental to the various places in the vicinity—Prideaux Haven, Roscoe Bay, Pendrell Sound, Toba Inlet and Refuge Cove—but the point is it is central to them all, and not far from Redonda Bay. For many years Redonda Bay was considered, as is Big Bay on Stuart Island today, to be the gateway to the north through the rapids of the Yucultas.

Toba Wildernest

Ed and Mary Schlote
Mouth of Toba Inlet at Double Island
Desolation Sound.
Ph (250) 286-8507
Surrey contact: (604) 597-5280

Dock available for overnight–100 feet. Facilities: Toilet, showers, cabins, ice. This is a fishing resort offering resort amenities for fly-in guests. Also available for transient mariners when space is available. It is located near Walsh Cove.

The south docks at Big Bay Marina. Boats to 130 feet can be accommodated on these and the adjoining dock. Right: The restaurant and store are just up the path.

Stuart Island

Big Bay Marina Resort

Bruce and Kay Knierim
Big Bay P.O. Stuart Island,
BC V0P 1V0
Phone: (250) 286-8107
Chart 3543

Hazard: Shoal in centre of Bay especially near public dock. Currents.

Marina services:

Fuel: Gas. diesel, oil.
Moorage, outboard motor repairs & service. Honda dealer.
Water at dock.
Power at docks: Limited: 15 amp.
Power at marina is being upgraded.
Fish cleaning station.

Customer services:

Restaurant, pub. Breakfast, lunch, dinner. Patio tables. Liquor store.
Store. Fishing licences, bait, tackle, ice, groceries, frozen foods, fresh bread, books, charts, gifts, video movie rental. Coffee, expresso, latte.
Laundry, Showers, Washrooms.
Public pay phones. Cottages.
Fishing charters. Guides. Good scuba diving nearby.

Entertainment.

Walk in coastal trails.
Games room at resort.

Adjacent facilities:

Pub, restaurant. Government dock.
Post office. Seaplane landing and mooring.

Big Bay lodge, marina and fishing camp, a place to stay on Stuart Island, refuel, replenish supplies, charter a fishing guide and much more.
There is moorage also at the nearby Wheelhouse pub or at the adjacent public dock.

Owen Bay

Transport Canada dock
Chart 3538, 3541
Manager •
Float length 9 m

Evans Bay,

Read Island
Transport Canada dock
Chart 3538, 3541
Float length 18 m

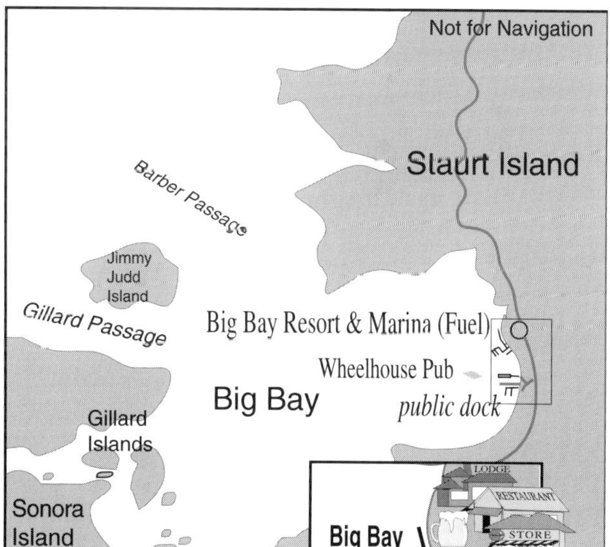

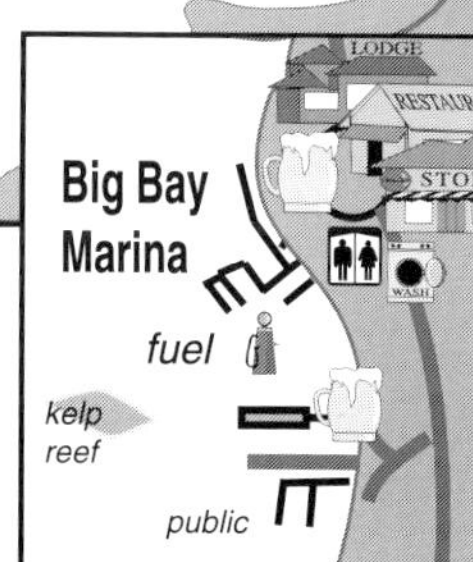

At Big Bay you will find fuel and moorage as well as an excellent dining facility and a large well stocked general store. The sport fishing out of Big Bay is outstanding. If you have the time lt the dock and include a day out fishing with a guide. Or you may simply sit on the patio and sip espresso and watch the coming and going of transient boats.

In Big Bay mariners are advised to enter and leave with caution, noting the shallows near the shore facilities. Consult your chart and watch for kelp–a summertime marker for the reef.

The perfect way to end your busy day in Big Bay is dining at the Eagle's Wing restaurant in the main building of the lodge. On cool days find a table near the warmth of the large wood-burning fireplace.

Shoal Bay

> Shoal Bay.
> East Thurlow Island
> Fisheries & Oceans dock
> Wharfinger: Nancy (& Chris) Carson
> Phone: "Loggers Dream" N114672
> Channel E. Thurlow 23.
> Float length 76 m. Chart 3543

Shoal Bay Lodge

Blair and Kevan McLean
P.O.Box 818 Campbell River,
BC. V9W 6Y4
Phone: (250) 286-6016 Fax 286-1936

Marina services:

Transient moorage. Government dock. Water available at dock from lodge–ask at Lodge for supply.
Pay moorage to wharfinger Nancy Carson

Chart 3543 VHF 23 radio phone
***Loggers Dream* N114672**

1998: Property for sale. Expect changes.

Customer services:

Ice. Restaurant–*Tracey Edwards*, chef. Dining. Patio service, licenced bar. **Laundry, showers, washrooms** (depends on water availability). Excellent fishing nearby. Accommodation.
Water taxi access. Flights.
John Hangodi and *Tracey* manage lodge.

Entertainment.

Hiking trails. Logging roads. View points along a network of paths. Animal and marine life. Eagles.

Adjacent facilities:

Canoeing. Kayaking. Day trips in adjacent waterways. Includes Thurston Bay Marine Park. Rentals. Picnic area.

At Shoal Bay a government dock with fairly generous moorage provides overnight accommodation for several large or additional smaller boats. The grounds of the lodge are pleasant for walking and the facilities include restaurant and store. The lodge offers meals and patio dining. The floating lodge below is in Cordero Channel–known internationally for fine dining and successful fishing charters.

Cordero Lodge

Doris & Kellie Kuppers
General Delivery
Blind Channel P.O.
BC. V0P 1B0
Phone (250) 286-8404
Chart 3544

Marina services:
Transient moorage.

Customer services:
Restaurant–Fine dining.
Lodging–up to eight guests.
Fishing: Boats–guided or unguided.

Entertainment.
Sunsets and views. Animal and marine life. Bears and eagles.

Adjacent facilities:
Cordero Islands anchoring.

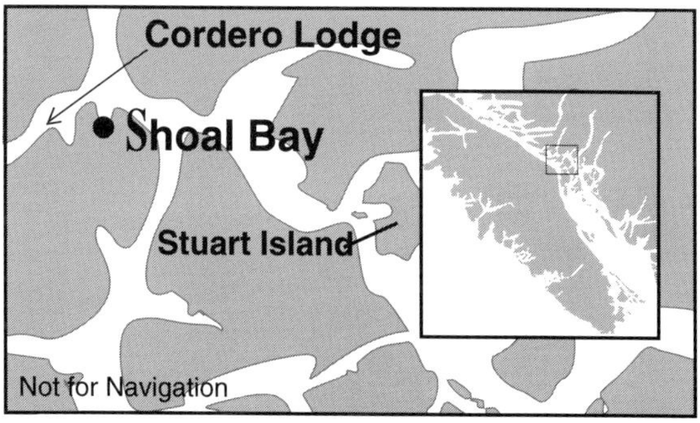

Blind Channel

Charts 3544, 3543 VHF 6 & 73

Blind Channel Resort

Philip Richter
Blind Channel BC
V0P 1B0
Phone (250) 830-8620, 286-8112
email: *richter@oberon.ark.com*
Web: *http://.oberon.ark.com/~richter*

Marina services:

Moorage: Transient moorage.
Fuel: Gas, diesel, propane.
Water at dock. Ice.
Power at docks: 20 amp

Customer services:

Restaurant open June—Labour Day. Fine dining-excellent cuisine. Patio. Groceries. Postal drop. Public phone (Satellite). Baked goods, bread, fresh produce, milk, frozen foods, ice, tackle, bait, charts. Fax service. **Laundry, showers, washrooms.** Liquor. Arts and crafts. Gifts.

Excellent fishing and prawning nearby. Scheduled flights.

Entertainment.

Hiking trails. Logging roads. Several view points along a network of paths. See the huge "Thurlow" Cedar tree. Incredible sunsets and views. Animal and marine life. Eagles. Dock are wide and embellishments are fascinating.

Adjacent facilities:

Picnic area.

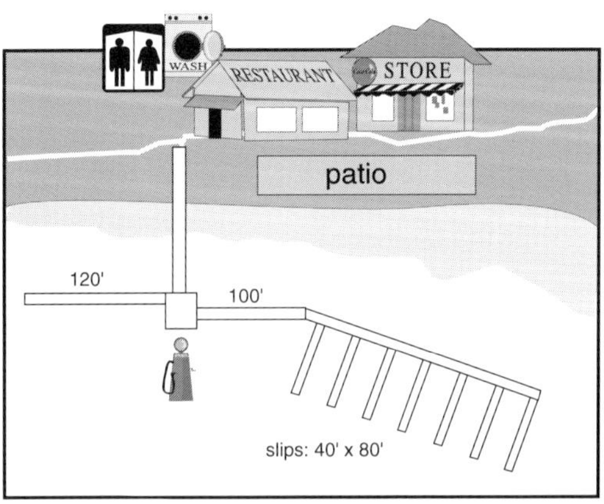

Blind Channel **Excellent Cuisine and more**

Three generations of Richters are busy around Blind Channel when guests are moored at their docks. This long-established marina in one of the coast's busy waterways has evolved over the years to become a well maintained and contemporary facility for cruising yachtsmen. Philip's parents, who established the marina still live and work at the property. Annemarie usually finds time to chat with guests in the restaurant while Edgar continues to be involved with planning and preparing further additions on the property. Philip does stints in the store while his wife, Jennifer, can be found preparing bread and other baked delectables for the store and restaurant. The marina offers moorage, fuel, water, propane and 110 volt shore power at the dock. The services include showers, laundry, ice, liquor agency, store and post office.

Nature lovers and hikers will be kept busy for days hiking trails that have been established by a large logging and sawmill company on West Thurlow Island. These trails are designed to show the features of a second growth forest and it just happens that they begin about 300 meters from Blind Channel Marina. There are three different trails, one to a spectacular viewpoint overlooking Mayne Passage and East Thurlow Island, a second to the Big Cedar, a tree with a diameter of 16 feet, via a forest of 80 year old second growth and the third through a thinned western hemlock stand that was naturally established in 1964. The final segment of this trail descends through 100 year old second growth.

The stop at Blind Channel has been a favourite among mariners for many decades and the Richter family continues to cater to their needs with fresh produce, baked goods and gourmet cuisine in a fine waterfront restaurant.

Across Cordero Channel, a few miles from Blind Channel is Cordero Lodge which also offers outstanding cuisine. It's a place to stop overnight, with limited space and usually requiring advance reservations for moorage and dinner. Owned and operated by Reinhart and Doris Kuppers, this popular facility has been a stopping place for those in the know for many years.

North to Port Hardy

Along the waterways of Queen Charlotte Strait, Kingcome Inlet and Tribune Channel I felt as though I could look down over the world of Vancouver Island and Desolation Sound far below. There the sets of turbulent rapids free the upper waters from their mountainous reservoir channels to flow and surge into the valleys of Desolation, Georgia Strait and ultimately Juan de Fuca as well as over the northern tip of Vancouver Island into the open Pacific. Currents running swiftly tug at kelp beds lining the shore, carry flotsam rapidly by, ripple and eddy, sometimes forming fascinating whirlpools.

Our first stop beyond Desolation was to have been Owen Bay in Okisollo Channel. But a slight miscalculation took us through Hole-in-the-Wall at more than half its full flooding speed of 9 knots, and so we headed into Octopus Marine Park for the night. And we were glad of our slight error. Octopus is an anchorage in another world—author

Many voyagers to the waters beyond Desolation Sound set out initially in fear of the awesome rapids: the Yacultus, Dent and Green Point Rapids. *Sailboating writers made me fearful of the rapids until the first time I navigated them.* Many boat owners never even plan to take in those waters during their years of cruising—perhaps because of a subconscious dread of the rapids and swirling narrows sometimes conjured up by the works of some experienced writers. But there is no need to fear the rapids north of Desolation Sound; just respect them. Respect them and work closely in conjunction with the tide and currents tables. With a hull-speed power boat, following the advice of sail boaters makes sense, but a fast powerboat will carry you through easily, when you would otherwise sit and wait in a slow boat. However, caution must be observed no matter what boat you have. Rapids and tide rips are dangerous and navigating fast flowing channels at maximum or near maximum currents is foolish and can be disasterous. I noted:

We passed through one set of rapids in the Yacultus area, stopped at Stuart Island, smack in between Yacultus and Dent, and waited for the following slack before continuing and passing through the second set. In between we hiked along the island trail to Arran Rapids to watch them run at full flood. (Unfortunately that trail is no longer accessible).

The Arran Rapids, one of the most carefully respected in the area, pour through a narrow channel and on a full ebb create a massive whirlpool at the southwestern end. I wrote: *It was a moderate flood that we saw, and no problem, boats large and small passed through in either direction. However, they were all high speed planing powerboats ranging from 13 ft Boston Whalers to 28 or 30 ft sportsfishermen. One fish boat with a 12 knot speed capability ran through with the flood with no apparent concern.*

The channels, straits and sounds beyond the Yacultus are a cruising yachtsman's paradise. They are relatively desolate, environmentally clean, aesthetically unspoilt and naturally quite undisturbed. There are fish, river otters, mink and eagles. We saw many, many bald eagles. And one gigantic golden eagle. There are dolphins, orcas, bears and deer. Some of these are quite elusive. My log entry:

Cruising out of Johnstone Strait at Alert Bay, it was convenient to transit Blackfish Sound, pass around Swanson Island into the entrance of Knight Inlet and off to port around Midsummer Island, up Retreat Passage to the most isolated, beautiful cove in the world at the north end of Bonwick Island. We anchored and set off in the dinghy to drop traps in the hopes of crab for dinner. Success. We went ashore and dug clams for chowder. And this prompted an urge to get on with some fishing. Spring salmon were running and in no time at all we hooked a couple of small ones — but eight pounds is plenty. We had prawns too—fished out of 45 fathoms off Minstrel Island.

Buildings at Karlukwees, a classic aboriginal native village on the coast of British Columbia. A totem lies near the water's edge at the Mamalilaculla village site.

We had intended to conclude our northward passage with a run up to the head of Kingcome Inlet of ***I Heard the Owl Call My Name*** *fame, but heavy, inclement weather kept us in Cypress Harbour opposite its mouth until time simply ran out and we had to begin the 'downhill' run back towards Vancouver. On a future trip we went up the river in an inflatable as far as Kingcome Village. But the time sitting out the weather was not wasted. We went out of Cypress to look for salmon and —the thrill of catching a big one is hard to equal.*

Indian villages are as exciting as the very names of the islands on which they are located or the channels that serve them. Arrow, Indian and Village Channels, Village, Owl and Midsummer Islands all conjure up romantic notions of history. The now deserted village of Mamililaculla (in the past few years members of the local band have cleaned up the ancient village and welcome vistors each summer for a fee towards the maintenance and upkeep of the property) and Karlukwees have about them an air of pressing history.

We made our way to Mamililaculla after anchoring near its disused jetty by hiking along the rocky shore to the beach in front of the deserted village. The path from the jetty has been cleared and can now be used to reach the historic site.

Top: Karlukwees and its disintegrating dock.
Centre: Looking down on the remnants of Karlukwees native village. Permission to go ashore at local villages must be obtained from the U'mista Cultural Centre in Alert Bay. (250) 974-5403.
Inset: Dock at Karlukwees. Bottom: Port Neville.

I wrote on my first visit to Mamaliliculla: *Sitting on a low ledge overlooking Native Anchorage, grey, weathered houses sit forlornly looking across the water. Long grass and weeds grow up around woodwork, porch steps, ancient longhouse frames. Totems, horizontal and fallen, slowly yield to the decay of time and elements. A gentle breeze flaps ragged cloth in window frames, or torn plastic used for temporary coverage since the village's abandonment—signs of recent overnight visitors.* Pathways, homes, community and mission halls were fast succumbing to the encroaching foliage. A sad confirmation of the prophetic accounts of Ms. Wiley Blanchet in her **Curve of Time.**

At Minstrel Island and in Lagoon Cove fuel, moorage and some services are available. In the upper reaches of the cruising waters farther north, there is fuel, a post office, store, arts and crafts and other facilities at Echo Bay. More services for mariners are available at Greenway Sound, Sullivan Bay and Shawl Bay near the mouth of Kingcome Inlet. Greenway Sound has an outstanding restaurant and boasts such luxuries as a hairdressing salon and a dog walk. Shawl Bay serves a pancake breakfast with overnight moorage.

The dock and old store (closed) at Port Neville. There is a post office ashore and an art gallery and gift shop on the opposite shore.

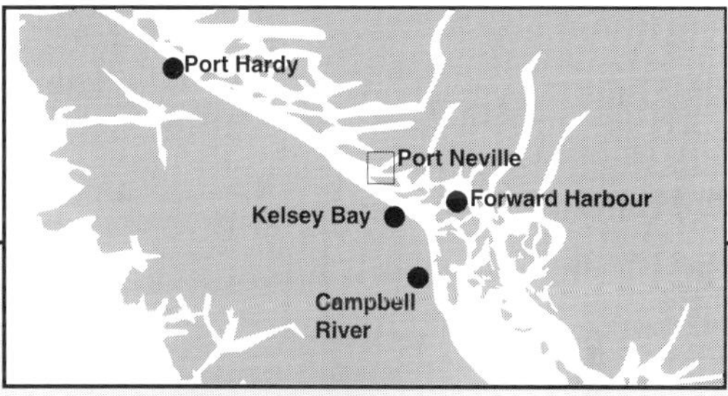

Kelsey Bay

Vancouver Island
Fisheries & Oceans dock
Phone (250) 282-3465
Chart 3544
Manager • Float length 177 m
Breakwater • Garbage • Water • Lights • Power •

Photo left: The late Olaf Hansen, his family settled at Port Neville in 1891.

Kelsey Bay

Vancouver Island
Transport Canada dock
Chart 3544
Manager • Float length 31 m
Breakwater • Water • Lights • Power •

Forward Harbour

Off Wellbore Channel (Whirlpool Rapids)
Chart 3544. See page 230

Port Neville Historic settlement.

Johnstone Strait
Transport Canada dock
Chart 3564, 3545
Address: Port Neville, BC. V0P 1M0
Call: *Sea Scout III.* VHF 6.
Float length 34 m • Managed by Lorna Chesluk Hansen (and Erica)–also post office manager. P.O open Mon, Wed, Fri. Forwarded mail held for arriving boats. Art gallery and gift store nearby.

Minstrel Island

Minstrel Island Resort

Grant Douglas and Sylvia Douglas
Minstrel Island P.O. 69,
B.C. V0P 1L0
Phone & fax (250) 949-0215
Chart 3564, 3545, 3515 VHF 73

Marina services:
Moorage: Transient moorage. (public)
Fuel: Gas, diesel, stove oil, propane, aviation fuel. **Water** at dock. Ice. **Power**.
12,000 lb dry dock–tidal grid.

Customer services:
Restaurant. Full menu. Speciality–Fish and Chips (reportedly "*awesome*").
Homemade bread, pies, jams. Pub. Accommodation in resort rooms and cottages.
Groceries. Postal drop. Post office now located nearby in Chatham Channel.
Cold beer. Bakery goods and fresh produce, milk, frozen foods, ice, tackle, bait, charts.
Boat rentals.
Laundry, Showers, Washrooms.
Good fishing and prawning nearby.

Entertainment.
Hiking trails. Incredible sunsets and views.
Animal and marine life. Bears and eagles.

Adjacent facilities:
Regular scheduled flights. Boardwalk.
Wheelchair access up ramp. Hiking trails.
Picnic area.
1998: Property for sale.

Forward Harbour Fishing Lodge

Al and Mary McAvoy (owners)
RR #3, Site 340, C-8
Courtenay, B.C. V9N 5M8
Phone (250) 268-8444
Chart 3544 VHF 73
Marina services: Water, Moorage.
Washrooms, Showers. Restaurant.
Fishing guides, hiking, walking.
Local interest: petroglyphs. Summer only.

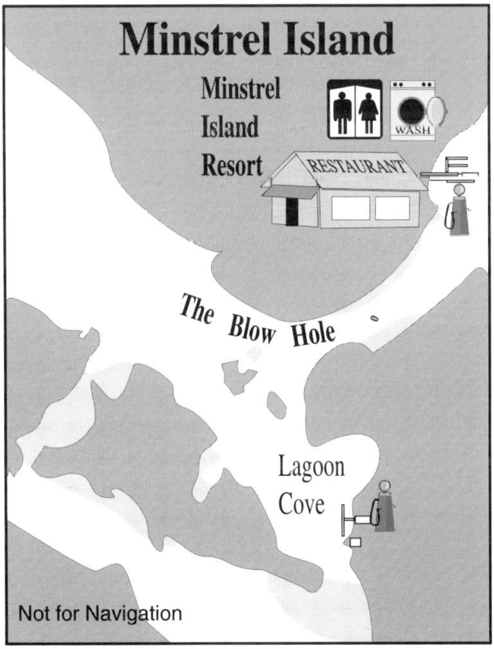

Minstrel Island

Fisheries & Oceans dock
Chart 3564, 3545, 3515
Manager • Float length 135 m
Aircraft float • Lights •
Managed by Minstrel Is Resort.
Adjacent facilities include: Fuel, restaurant, lodge, store.

The marina at Minstrel Island is protected from most wind and weather. The public dock is adjacent to the private property of Minstrel Island Resort and is subject to government maintenance and regulations. They provide access to the shore where Fish and Chips are the speciality at the full menu restaurant.

Lagoon Cove

Lagoon Cove Marina

Bill and Jean Barber
c/o Minstrel Island P.O.
British Columbia.
Chart 3545, 3564. VHF 73
post office now located in Chatham Channel

Marina services:

Moorage: Transient moorage.
Fuel: Gas, diesel, propane. Oils.
Some repairs subject to available help.
Haulouts. Charts, books.
Water at dock. **Power:** 15, 20 amps.

Customer services:

Coffee kiosk. Fishing licences, ice, tackle, bait, sodas, candies. Crab and prawn traps.
Showers, washrooms.
Boat rentals.
Excellent fishing and prawning nearby.
Cabin and boat available to fly-in guests.

Entertainment.

Hiking trails. Sunsets and views. Animal and marine life. Bears and eagles.

Many people have chosen to anchor in Lagoon Cove over the past, but with secure docking and the associated peace of mind when the wind is up it is worth while stopping at the marina. There are showers, washrooms, additional accommodation and the friendly greeting of owners, Bill and Jean Barber.

Former owners accumulated an incredible collection of marine and other hardware, enough for a museum. Some items have been discarded but many remain in a work shed display area and patio coffee nook.

Double Bay Resort

Double Bay Resort on Hanson Island can be seen to the left in Double Bay.

Double Bay Resort

Murray, Cheri and Shawna Gardham
371 McCarthy Street,
Campbell River BC. V9W 2R7
Phone: (250) 949-2500 or 286-1937
Chart 3546 VHF 73
Web: *www.doublebayresort.com*
Moorage available on 700 foot dock.
No dock services. **Showers, washrooms.** bait, ice, tackle. Restaurant 8am to 8pm–home cooked meals. Fishing resort offers fishing charters. Rental boats.
Open May–September.
View eagles, whales, dolphins.

Stubbs Island Charters

Jim Borrowman,
P.O. Box 7, Telegraph Cove,
BC. V0N 3J0
Phone: (250) 928-3185 Fax 928-3102
Chart 3546 VHF 10
email: *stubbs@island.net*
Web: *www.stubbs-island.com*
Customer Services:
Gifts. Fine art. Arts and crafts. Whale watching, day trips. Scuba diving charters.Accommodation. Cabins on boardwalk.

In the cove a new development on the adjacent shore includes docks marked "No Moorage". Like those already in place at the adjoining marina these docks are built for small boats making it very tight for medium to larger vessels to squeeze into the cove. Stop at the fuel dock and visit the craft shop on the wharf. The fuel dock does not belong to the store on the wharf. It is operated by the marina farther in the cove. There is fuel also at Alert Bay and Port McNeill.

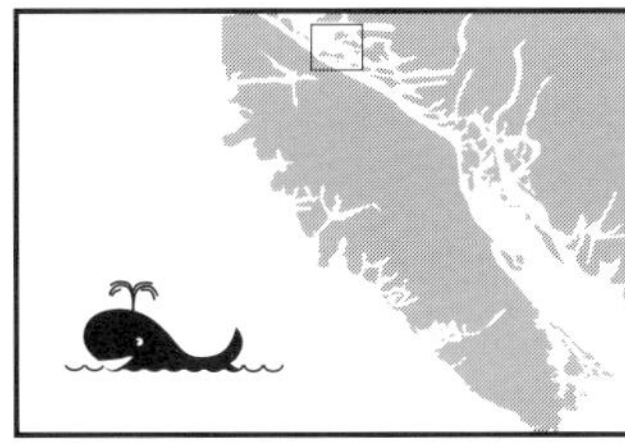

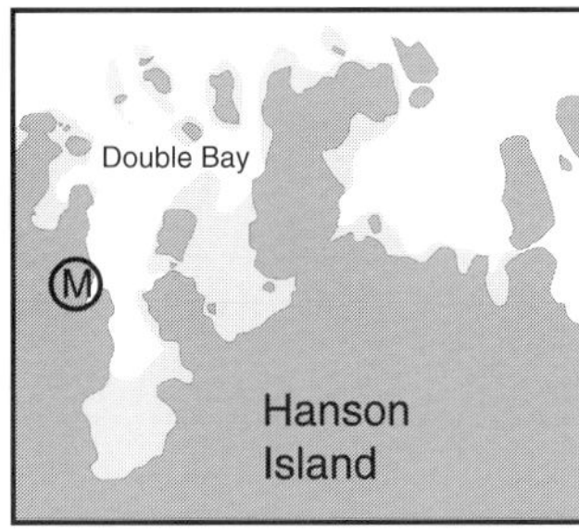

Photo left: Fuel dock at Telegraph Cove.
Lower: The launch ramp inside the cove.
Below: Double Bay—shows marina tucked inside.

Telegraph Cove Resort

Telegraph Cove, BC. V0N 3J0
Ph (250) 928-3131 Chart 3546
Moorage: Moorage mostly reserved for resort clients.
Fuel: gas, oil. **Water. Launch ramp.**

Customer Services:
Fishing resort, accommodations, camping, RV parking.
Groceries, bait, ice, snacks.
Fishing tackle, licences. Public phone.

Entertainment:
Scuba diving and whale watching–
Contact Stubbs Island Charters. Marine and bird life. Restaurant.

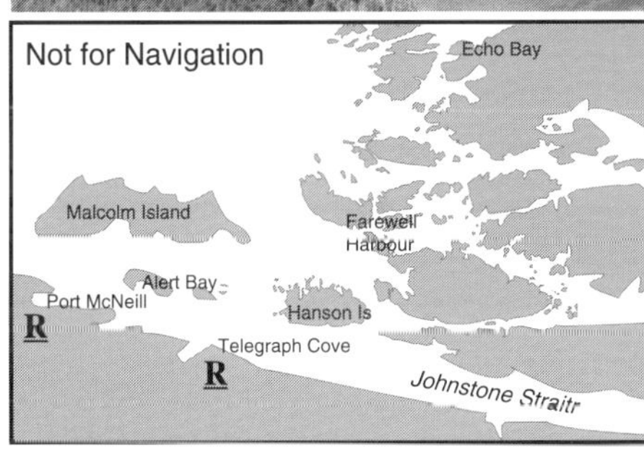

Native Villages, whales, people and places

Telegraph Cove is one of the most picturesque spots on the BC coast. Each of its quaint buildings, which stand on a boardwalk supported by ageing pilings, has a story of its own and today reflects life on the coast in the earlier part of this century. A steady flow of non-boating visitors each year descends on the village to set out on whale watching or scuba diving excursions to locations in and around Blackfish Sound. Mariners preferring to keep away from the hustle and bustle of activity in places like Telegraph Cove head for the numerous tranquil anchorages that abound, or the historic Indian villages that have been immortalized by writers and visitors in past decades. Mamaliliculla and Karlukwees are nearby native villages which I recall visiting in the mid seventies and at which time they appeared to have been long since abandoned. Today I realise they had not been abandoned that long ago, especially considering the decay and reclamation of the undergrowth and vegetation that has taken place between my earlier visits and more recent years.

Telegraph Cove was little more than a sawmill for many years since its settlement by Mr A.M. Wastell decades ago. In the early part of this century salmon fishing prompted the construction of a saltery at the settlement. Other industries grew from the lumber and fishing interests, attracting labour and settlers and turning the sawmill settlement into a coastal village. In 1911 the federal government installed a telephone station at the cove to connect the north end of Vancouver Island with Campbell River. The second world war brought the cove's telephone and telegraph cable system into operation as a communications relay centre. For many years accessible only by water, today an unpaved road runs to the cove from the main Island Highway connecting the northern towns of Port McNeill and Port Hardy to the rest of the world.

Remnants of bygone activity and population in Bones Bay near Minstrel Island.
A quiet anchorage in a corner of Potts Lagoon.

In more recent times tourism has played a large role in the significance of Telegraph Cove. The revelation that some of the world's best scuba diving is possible within a short boat ride of the cove has brought many visitors from all parts of the world to view the abundant marine life at nearby sites such as Stubbs Island and Plumper Rock. Among the diving enthusiasts to have visited the area are Jacques Cousteau and underwater cinematographers Al Giddings and Stan Waterman who were among those responsible for the subsea footage in James Bond movies, The Deep, Jaws and many more. National Geographic Magazine has done major editorial features on the area which included large underwater colour photo spreads by staff photographer David Doubilet. Working with these personalities were individuals such as Jim Borrowman and former associate Bill McKay. Jim runs a charter boat service catering to sightseers, whale watching groups and divers. Whale watching is one of the major industries at Telegraph Cove. People travel from all areas to see killer whales in the wild, and Jim and his staff not only take visitors to see whales but also offer an interesting and comprehensive running commentary on the habits, histories and behaviour of the whales.

Telegraph Cove includes a compact, busy marina with a store and parking lot for patrons in a compound adjacent to the ring of private houses on pilings at the water's edge. At the entrance to the cove from the upper central reaches of Johnstone Strait a small dock offers brief stops to take on fuel. Fuel stops with more generous facilities are located at Port McNeill and Alert Bay. Just at the top of the ramp is a post office and store where souvenirs can be purchased or where arrangements can be made for sightseeing, whale watching and scuba diving. The whale watching/scuba diving operators also offer accommodations in several of the old houses which have been refurbished and equipped for guests. Summertime is a busy period for whale watching with the season peaking through July and August. In winter, spring and fall scuba diving comes into its own and regular groups of divers make use of the facilities and the two whale watching boats to access the legendary dive sites nearby.

By private boat, charter boat, excursion or whatever other means, there are many outstanding attractions not far from Telegraph Cove. These include beautiful moorings and anchorages in the nearby island groups such as the Plumpers on the edge of Weynton Passage, access ashore at abandoned and historic native villages and coastal settlements with excellent fishing, wildlife viewing and just simply cruising among some of the most beautiful of the BC coast islands, inlets and waterways.

When anchoring overnight in nearby sheltered waters be careful to check on the effects of tide, current and wind before settling in. It is tempting to anchor in the Plumper Islets. A small cove is formed between two of the larger islets which almost join at the northern end. A local diving club has moored a raft in the middle of the cove and this appears suitable for tying alongside. At higher tides and during tidal changes the water tends to rip through the northern passage causing quite a current through the Plumpers. If you tie up to the raft use the right side entering the cove from the direction of Telegraph because the current is milder than on the left. Or anchor between the raft and the island to starboard.

Make sure to stern tie as well. Unless winds accompanied by high tides are expected this may be a secure overnight anchorage. It affords spectacular sunset views across towards Telegraph Cove. There are many other anchorages that are larger and safer nearby, specifically in the back of Hanson Island, Harbledown Island and up Baronet Passage at places like Potts Lagoon, Bend Island and Clio Channel. See the companion guide: ***Anchorages and Marine Parks.***

Now you are in historic native Indian territory. Visting the deserted villages of Karlukwees and Mamalilaculla is just a short distance from nearby anchorages. The villages have temporary anchorages or dilapidated dock space and overnight stays are not generally recommended. Go ashore, look around but leave everything as you found it. Nearby Minstrel Island offers safe and secure overnight moorage and the opportunity to replenish fuel, groceries and provisions to continue your cruise. Access to these areas from Telegraph Cove is easy and close, but be careful with your navigation as some passages are narrow and strewn with rocks and reefs. Use a large scale chart.

The Indian villages of Mamalilaculla and Karlukwees have long been a favourite stop among cruising yachtsmen. Often shrouded in swirling fog, drenching drizzle or heavy cloudy overcast, an atmosphere of history and the presence of spirits of the past lingers in these villages, pervading the ambience of the shamble of houses, tangle of bushes and decayed ritualistic and ceremonial structures. A tilted totem pole peeked jauntily from between the branches of the heavy bushes on a previous visit.

Huge longhouse beams and several horizontal totems fought off the encroaching grasses and weeds, brambles and berry bushes. When Wylie Blanchet (**The Curve Of Time**) visited the area in the 1920s there were people living there and she mused over her prediction that it would not be long before the villages would be abandoned. She was right. But today visitors will find efforts in place to restore historic testimony to what was once a thriving community.

It's a pity that historic Minstrel Island has long since seen the last minstrel show in its grand community hall that once served as an entertainment centre for the area with patrons coming in from nearby Bones Bay (cannery) or from logging installations and camps up Knight Inlet or adjacent waterways.

One of my favourite sightseeing excursions in the area is to Alert Bay which should include a hike to the spectacular and almost unheard of *Gator Gardens* located atop the hill behind the radio station. It is reached by a signposted and fairly easy walk up the back roads.

Going in the opposite direction out of Telegraph Cove is Robson Bight, on the west side of central Johnstone Strait. It is here that killer whales congregate in summer and where mating takes place. For years boats would chase the orcas around this area for close views, or stand drifting in the bight for the occasion when the whales would inquisitively come up to the boats and give the crews a close encounter. Chasing of orcas became a problem and in recent years controls have been

The public dock at the far end of Alert Bay

applied to limit the access by boats to the habitat of the killer whale population. Cruising anywhere in the vicinity of Telegraph Cove, upper Johnstone Strait, Blackfish Sound and adjacent waters, could present an opportunity to see killer whales in the wild. Eagles abound too. Often you will find the large birds sitting high in the trees overlooking Baronet Passage, especially on Cracroft Point just off Johnstone Strait where many fishermen try for big salmon in the current swept pass. The eagles will select a boat in what they seem to claim as 'their' territory and when an undersized or unwanted fish such as a quillback or other rockfish is thrown back the eagles will swoop down to claim it as their prize. Other eagles tend to respect their neighbour's territory and not dash in to compete for the free meal, as perhaps would the seagull population.

Many, if not most, places become lost to history. Until recently, thankfully quaint, picturesque Telegraph Cove remained standing as a thriving community, centre for whale watching, scuba diving and sportsfishing and most importantly as a classic historic site representing a most colourful and important segment of British Columbia's past. New development seen on the bluff as you enter the cove, along with new buildings and 'improvements' at the marina appear to be changing all of that.

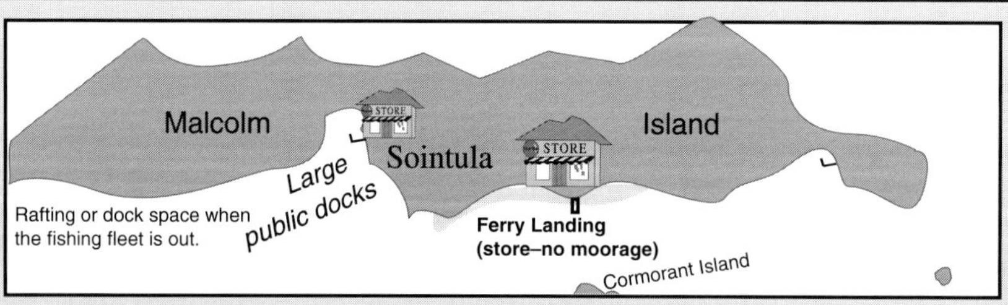

Sointula

Malcolm Island Lions Club
Harbour Authority
Phone (250) 973-6544
Chart 3546
Manager • Float length 745 m
Breakwater • Aircraft float

Malcolm Island Beachcombers Inn
Phone: (250) 973-6366
• Rooms • Showers • Garbage • Water • Lights • Power • Public phone • Store
Nearby facilities. Road access to ferry to Port McNeill, Alert Bay.

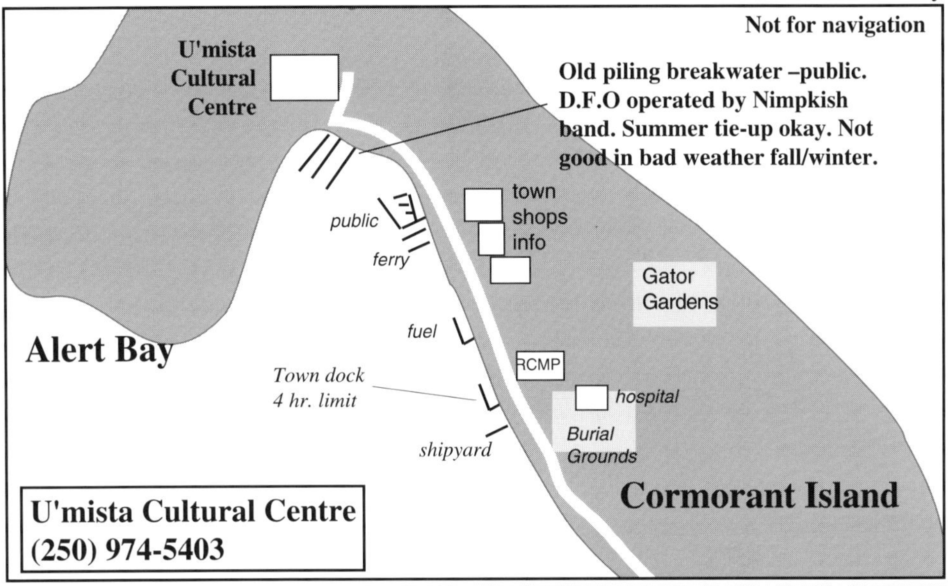

Alert Bay

Alert Bay Boat Harbour

(250) 974-5251 res. 974-5727 office.
Government dock.
Chart 3546 VHF 73
DFO (Fisheries) VHF 78

Marina services:

Transient moorage.
Power. Water.

Customer services:

Restaurants. Dining.
Groceries. Tackle, bait, charts.
Garbage disposal.
All services at village of Alert Bay.
Accommodations, meals.
Credit Union (bank machine). 974-5527.

Entertainment.

Hiking trails and roads. **Gator Gardens**, an anachronism–Florida's Everglades come to BC. On the hill above the town. Animal and marine life. Prime scuba diving nearby. **U'mista Cultural Centre.** Visit this outstanding display of native history.

Nearby facilities:

Paved airstrip (2800 feet).
Marine services: Alert Bay Save On Fuels: (250) 974-5411. Alert Bay Shipyards: (250) 974-5446. Alert Bay Travel Info Centre: (250) 974-5213

ALERT BAY

Fisheries & Oceans dock
Phone (250) 974-5727 res: 974-5251.
Chart 3546
Manager • Float length 533 m
Breakwater • Aircraft Float • Garbage
Water • Waste oil service *Best to return used oil to fuel stations* • Lights • Power • Public phone • Walking–*Gator Gardens* on the hill. Visit *U'mista Centre.*

Transport Canada dock
Chart 3546
Manager • Float length 61 m
Launch ramp •
Garbage • Water (on wharf only) •
Lights •

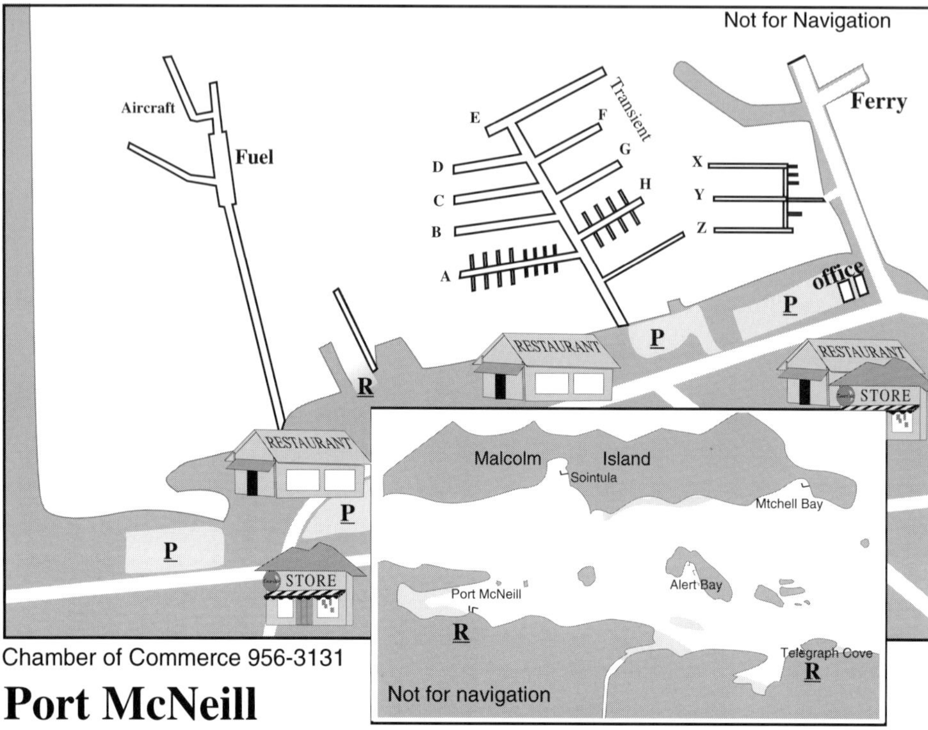

Chamber of Commerce 956-3131

Port McNeill Boat Harbour

P.O. Box 1389, Port McNeill, BC. V0N 2R0 Phone: (250) 956-3881
Harbour Manager
Chart 3546, 3548 VHF 73
Moorage: Extensive sheltered docks. Outer floats for transient moorage. Oil disposal. **Water. Power**: 50, 30, 15 amps. **Fuel** (adjacent)**:** Gas, diesel. Propane. Oils. Aviation fuels. **Launch Ramp** adjacent.

Services:

Garbage disposal. Tidal grid. Fish cleaning station.

Entertainment:

Walk along shore and sea wall. Uptown facilities–hotels, restaurants, shops, Heritage buildings. Various services. Vehicle rentals. Flights.

Adjacent facilities:

Ferry to Sointula, Alert Bay.
Marine repairs–adjacent and nearby Marinas/fuel barge. Also offering ice, showers, snacks and pop, hospital nearby.

R = Launch ramp
P = Parking lot

Opposite, top: Port McNeill. Inset: Well known gateway to burial ground at Alert Bay. Bottom: The lesser-known Gator Gardens on the hill near the radio tower.

Alert Bay

Alert Bay attracts mariners as a stop for replenishment and an opportunity to go ashore for some exercise as they travel en route to points north or home again. For many Alert Bay is a final destination on their northward travels. Whatever the reason for stopping at Alert Bay surprisingly few people seem to know of the existence of Gator Gardens. This very name may well indicate that this attraction is somewhat misplaced here in British Columbia. In name and character it is more like something you would expect to stumble across on a tour of Florida. It is a marshy, swampy

glade complete with large still pools of water afloat with the massive leaves of various forms of vegetation and sprouting large sprays of skunk cabbage. Wooden walkways have been erected across the park to allow access for easy walking and viewing. The most prominent feature of the park is its incredible trees which appear to have been struck by lightning at one time. These massive trees are broad and tall and mostly scarred and craggy with eerie looking branches and cracks and splits appearing as though they were the inspiration for the tale of Sleepy Hollow. And the surprising thing is that the entire park is not where you would expect to find it, down near sea level, but rather up on top of the hill overlooking Alert Bay. From the government marinas of Alert Bay to Gator Gardens is a good uphill walk to the back of the residential area overlooking the bay. There are several routes, marked here and there by signs indicating the way. All routes end up alongside or near the transmitter station and entrances to the Gardens, although not well marked, will eventually lead you to the wooden planked walk through the glades.

Alert Bay boasts the once tallest totem pole in the world. It is located a short walk up the hill behind the U'umista Native Cultural Centre which is on the shore adjacent to the government docks north of town.The museum at the cultural centre is well worth a visit. Among other interesting items it has on display segments of the exhibits that were shown at Expo 86 in Vancouver.

You may be lucky when visiting Alert Bay and experience some rare calm, sunny weather. If not you should watch the currents and sea conditions that sweep around the northern channel en route to Port Hardy or the open northern reaches of Johnstone Strait. In windy conditions it is usually possible to sneak around the bottom end of Malcolm Island and through the rocky channels and islets in the area. We once took shelter for two days in the Plumper Islets while the wind raged. It is not really a suitable anchorage for more than a temporary stop because the current rips through quite fast especially at high tides.

Echo Bay

Echo Bay Resort

Bob & Nancy Richter

Simoom Sound P.O.

B.C. V0P 1S0

Phone and fax (250) 956-2121

email: echobay@island.net

website: www.echobayresort.com

Marina services:

Transient moorage. Reserve in peak season–July, August. Open three days per week in winter.

Fuel: Gas, diesel, kerosene, oils, propane. **Water** at dock. Ice. **Power**: 30, 15 amps. 12,000 lb dry dock. Can haul out boats to about 23 feet.

Laundry, showers, washrooms (for overnight guests.)

Customer services:

Use Interac/Debit card.

Check your email at office. Groceries. Post office. Public phone. Bakery goods and

Charts 3515 VHF Ch 73

fresh produce, milk, frozen foods, ice, tackle, books, charts, film, and gifts.

Eight lodging units. Hotel for groups of six or more. Boat and motor rentals.

Popular fishing and prawning nearby

Block and party ice.

Entertainment.

The main dock is a part of the former Lake Washington floating bridge. Hiking trail and park access. Incredible sunsets and views. Animal and marine life. July 8 to August 23–Wednesday nights: Potluck dinners. Resort supplies fish or ribs.

Adjacent facilities:

Regular scheduled flights. Local arts and crafts shops and artists. Nearby anchorage and public float. Marine park.

Note: Docks have had an upgrade to supplied electrical power. Other improvements include new dock ramp to land.

Above: View looking out of Echo Bay with Echo Bay Marina on the left, the public dock in the foreground. The marine park and public dock are located alongside a delightful, white, sandy beach.
The aerial photograph opposite shows Echo Bay Resort on the right of the bay.

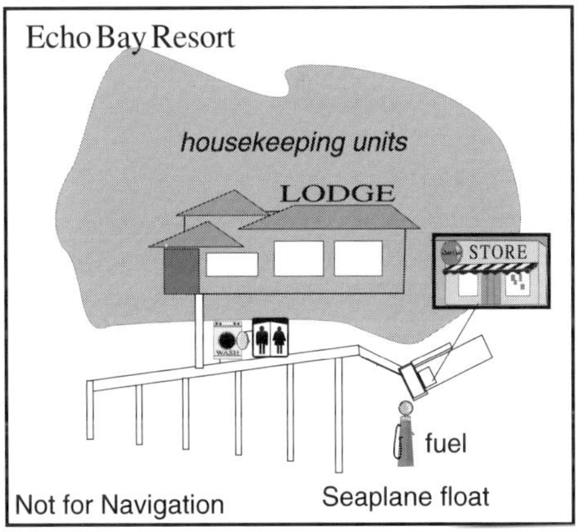

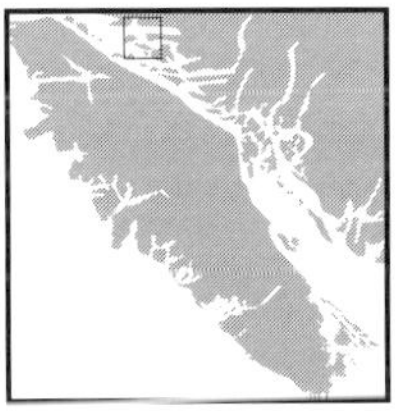

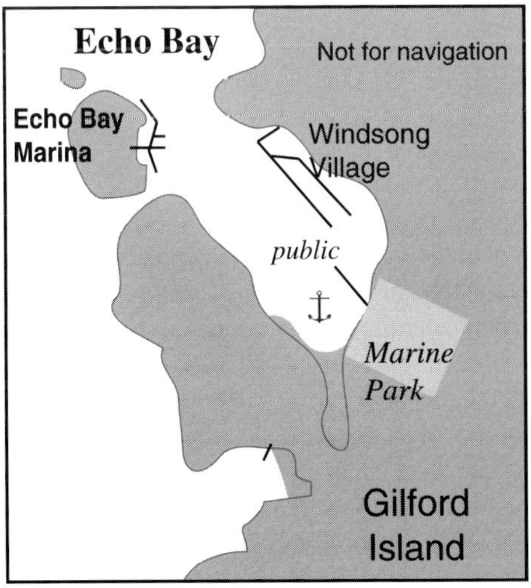

Echo Bay

History and commerce converge at Echo Bay. It is a hub of activity drawing summertime travelling boaters, whale watching tourists out of Telegraph Cove, sportfishing groups from near and far and a constant flow of local people from neighbouring logging camps and fish farms to pick up and drop off their mail and replenish some of their grocery needs. The lodge on the island at the entrance to the cove, for that's all that Echo Bay is—a tiny sheltered cove, caters to a steady flow of itinerant visitors in for a few days of fishing or stopping by in their boats for a spell in the area. Fuel and moorage with

Kingcome

Transport Canada
Chart 3515
Float length 55 m
Windy and exposed.

Gilford Island

Fisheries & Oceans
Chart 3515
Float length 60 m
Aircraft float •

Echo Bay (Marine Park)

All weather anchorage
Boat dock, camping sites, water, toilets..
Waddington Bay
Part of Broughton Archipelago (Park)
See *Anchorages and Marine Parks*

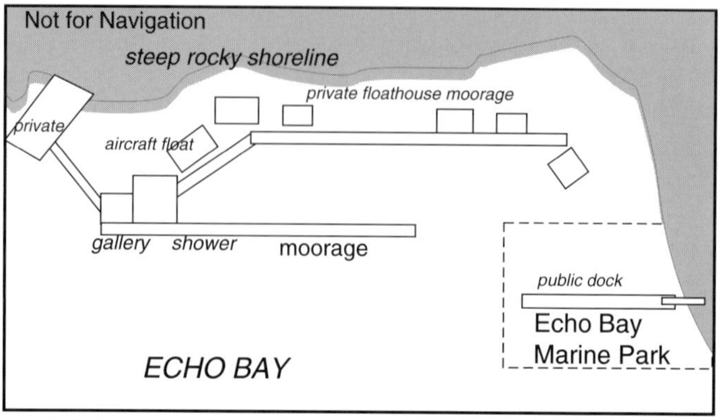

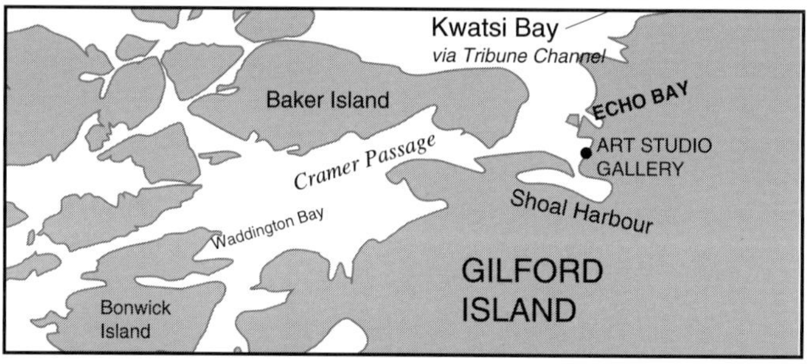

Right: Maximchuk gallery and art studio just outside Shoal Harbour entrance. Above right: Diagram and photograph (opposite page) show the Windsong Sea Village with its spacious docks, rental cottages and arts and crafts shop.

power and water are available from Echo Bay Marina which also serves as post office and store.

A marina on the opposite shore, Windsong Sea Village Resort, has large floats which will accommodate a good number of boats and offers moorage for seaplanes too. This marina offers no fuel or dockside electricity but has a delightful craft shop representing works of local artists such as Alexandra Morton, Yvonne Maximchuck, natives from the nearby Gilford Island Village and the proprietor herself, Christine O'Donnell.

Yvonne Maximchuck has a house just outside the entrance to Shoal Harbour and provides an opportunity for artists or tyros to brush up on their skills. Here, in her gallery studio, she conducts instruction and classes in art, water colour, acrylics and oils, available to people on visiting boats. She also has a fine selection of works for sale and can be reached by calling her at *Sea Rose* on channel 16.

Windsong Sea Villag e Echo Bay.

Christine O'Donnell
Box 388, Port McNeill
B.C. V0N 1A0
Phone: (250) 974-5004
Ph/Fax (250) 974-3009

Marina services:

Moorage: Transient moorage. Float planes. Showers.

Customer services:

Arts and crafts store. Fine art. Floating cabin rentals. Floathouse moorage available.

Entertainment.

Hiking trail and marine park access.
Incredible sunsets and views.
Animal and marine life.

Adjacent facilities:

Regular scheduled flights. Marine park and trail. Local artists live in vicinity. Their work is available. Nearby anchorage and public float.

Chart 3515 VHF Ch 73

Kwatsi Bay Ventures

Doug and Anca Knierim
Simoom Sound, V0P 1S0
Cell (250) 949-1384 Chart 3515

Moorage. New floats and facilities under construction.
Water (plentiful, good). Nearest fuel at Minstrel Island, Lagoon Cove, Echo Bay or Sullivan Bay.
This is a new marina at remote Kwatsi Bay off Tribune Channel. Access from Echo Bay, or en route from Minstrel Island. Beautiful scenery in Tribune Channel area. Ask about rock climbing.

Below: Kwatsi Bay lies beneath towering mountains.

Shawl Bay

Chart 3515 VHF 73

Brown's Marina, Shawl Bay

Lorne and Shawn Brown
General Delivery
Simoom Sound P.O.
British Columbia, V0P 1S0.

Marina services:

Moorage: Transient moorage.
Some provisions.
Power, water.

Entertainment.

Animal and marine life. Picnic float.
Pancake breakfast included with moorage.
Visit Kingcome native village nearby.

Adjacent facilities:

Water in good supply but use sparingly during summer. *Look for the turquoise docks to locate the marina in the bay.*

Shawl Bay

Shawl Bay is a busy place during the summer season. Many regular boating friends and customers of the people who own and operate the cosy marina at the southern corner of the bay return each year or periodically to tie up at the spacious docks. These docks and the structures on them comprise the Brown's Marina operated until 1994/5 by the late Edna Brown and her sister Johanne along with Edna's son Gary. Brother Alf Didriksen ran a logging camp in the bay for many years and the family's hospitality is legendary among fishermen, loggers and pleasure boats alike. It is now being run by Lorne Brown his wife Shawn and aunt Jo.

A large floating platform was cleared for moorage customers to come ashore and relax, socialize and perhaps set up their own barbecue or picnic. Ask about excursions to Kingcome Indian village.

Opposite: Inset photos show pancake breakfast at the marina–comes with the moorage compliments of the owners.

The floats and store on the dock at Shawl Bay. Aerial photo shows location of the marina at the far corner of the bay.

Greenway Sound

In 1985 Tom and Ann Taylor opened their floating resort at nearby Greenway Sound. The bay where their lodge is located is large and the docks extensive totalling one half kilometer in total length. They are laid out in a wide square 'u' with a finger pointing into the square near the far corner where the facilities are located. These facilities set the floating resort apart from others in the area. The store is part of a restaurant with an airy, clean kitchen that serves up mouth-watering dinners, wholesome lunches—hamburgers, chili, sandwiches, salads, excellent clam chowder and ice cream in waffle cones— and splendid breakfasts.

The moorage is typically busy with large visiting yachts, especially from the USA, some of which remain during summer and have owners and their friends fly in for brief or extended visits.The resort literally gives arriving boats the red carpet treatment, from the smiling attendance of young dock helpers to the full length coverage of the floats with red indoor/outdoor carpet. It is not unusual to see boat crews walking the docks for their daily exercise. Walk the length of the floats twice and you have walked a mile. Another walk that is being favoured nearby is an area ashore that has been designated as parkland by the Ministry of Forests. A dinghy float and ramp are being installed jointly by the Taylors and the Ministry Moorage customers or anyone who calls in the area will have access to the park. A path from the ramp leads to the lake a short distance up the mountainside.

The resort offers fly in service from points south including Seattle (Tacoma, Renton, Anacortes), Campbell River and Vancouver. The store carries produce including fresh and frozen foods, books, charts, tee-shirts and hats. In separate buildings attached to the adjacent docks are laundry and shower facilities as well as a hair and beauty salon and a book exchange library. The docks are serviced with water and 120/208 volt shore power. The resort stands by on channel 73 and is equipped with regular telephone service. It is possible to reach Alert Bay on VHF 86 for BC Tel radio operated service, and cellular is within close reach with the likelihood that it will provide full coverage at the dock soon.

Greenway Sound

Charts 3547, 3001 VHF Ch 73

Greenway Sound Marine Resort

Tom and Ann Taylor
P.O. Box 759 Port McNeill, BC.
V0N 2R0 (Also at 19924 Aurora Ave. North #54 Seattle Wa. 98133–off season)
1 800-800-2080 Ch 73
(250) 949-2525 (Port Hardy)

Marina services:

Moorage: Large permanent marina with plenty of transient moorage at up to 2,700 feet of red carpeted dock. Good for power walking. Reserve dock space in peak season. **Garbage** disposal (for mooring customers).

Water at dock.

Power at docks: 50, 30, 15 amp. 120/208 shore power available.

Facilities:

Restaurant. (Reservations essential). All meals. Breakfast at 11 am (lunch), dinner 6:30-8 pm. Licenced. Food and beverages.

Groceries. Baked goods and fresh produce, milk, frozen foods, ice, tackle, books, charts, film, and gifts. Video movie rentals.

Laundry, showers, washrooms.

Postal service and phone available.

Boat sitting.

Excellent fishing and prawning nearby.

Block and party ice.

Hair/beauty salon, barber shop.

Entertainment.

Take-out food–pizzas, baked goods, snacks. Book exchange. Hiking trail and lake access. Incredible sunsets and views. Animal and marine life.

Customer services:

Regular scheduled flights. Assisted travel arrangements. Tacoma, Renton,(Sea-Tac) Anacortes departures. *Greenway Sounder. Plus Kenmore Air and Air Rainbow.*

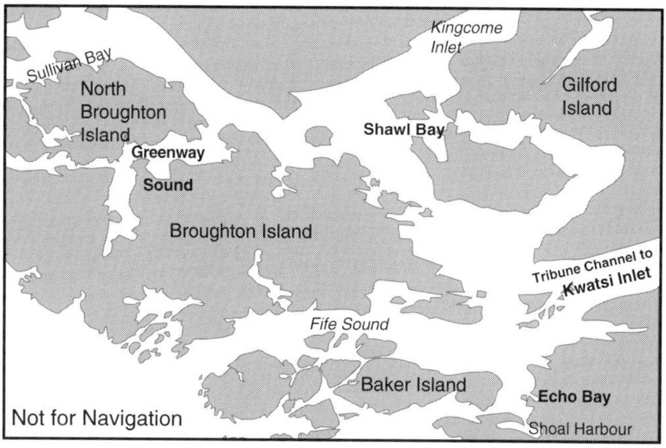

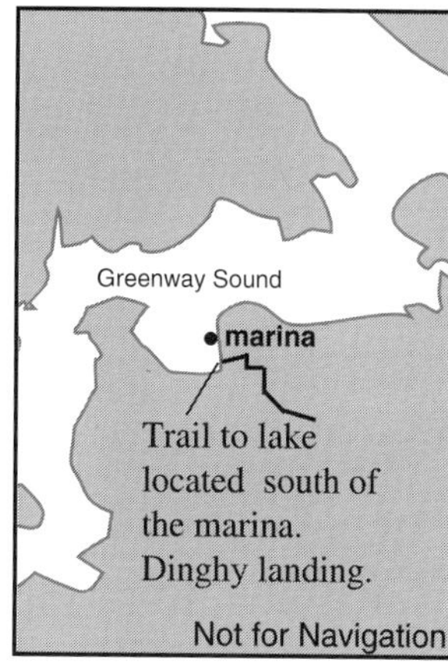

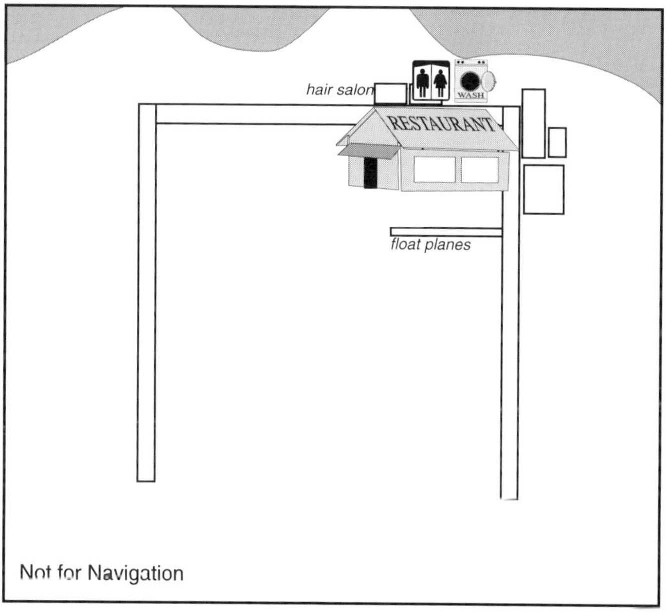

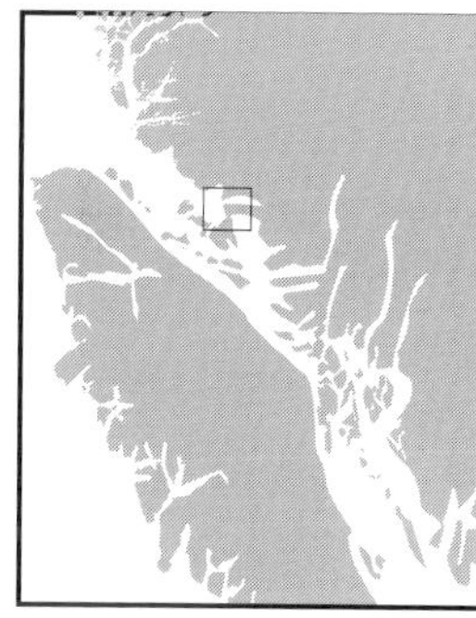

The restaurant is on a float alongside the dock. It also houses a small store as well as the marina office. An aircraft float has been added to the far dock since the aerial photograph was taken.

Opposite: An aerial view of Greenway Sound Marine Resort. Nearby, a hiking trail leading to a lake has been established. It is being maintained by the resort in conjunction with the BC Forest Service that provided the trail. Right: Restaurant and marina.

Sullivan Bay

Sullivan Bay Marine Resort

Pat Finnerty and Lynn Whitehead
Sullivan Bay, BC. V0N 3H0
Phone: (250) 949-2550

Marina services:

Fuel: Gas, diesel, oil, ice, bait.
Mechanic and services at Vancouver Island towns and ports. Some repairs possible.
Moorage: Large permanent marina with plenty of transient moorage at up to 4,000 feet of dock. Reserve in peak season.
Water at dock. Plentiful.
Power at docks: 50 , 30, 15 amp.
110 plus 220 shore power available.
Laundry, showers, washrooms.

Customer services:

Liquor store. Groceries. Post office. Fresh produce, dairy products, frozen foods, ice, tackle, and souvenirs. Video movie rentals. Smokehouse.

Chart 3547 VHF Ch 73 and 16

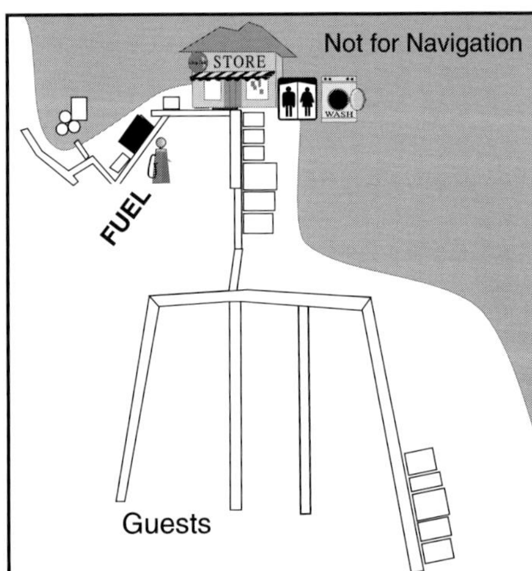

Boat sitting.
Fishing and prawning in vicinity.

Entertainment.

Library, TV lounge. Novel building structures, street names on docks.

Adjacent facilities:

Regular scheduled flights. Private floating homes village.

Views of Sullivan Bay Marine Resort with its floating sidewalks, permanent moorage, client residences and itinerant visitors.

Sullivan Bay has long been known for its remote but popular location on the BC coast. It has served for many years as an air traveller's focal point and as a final destination or layover point for mariners.

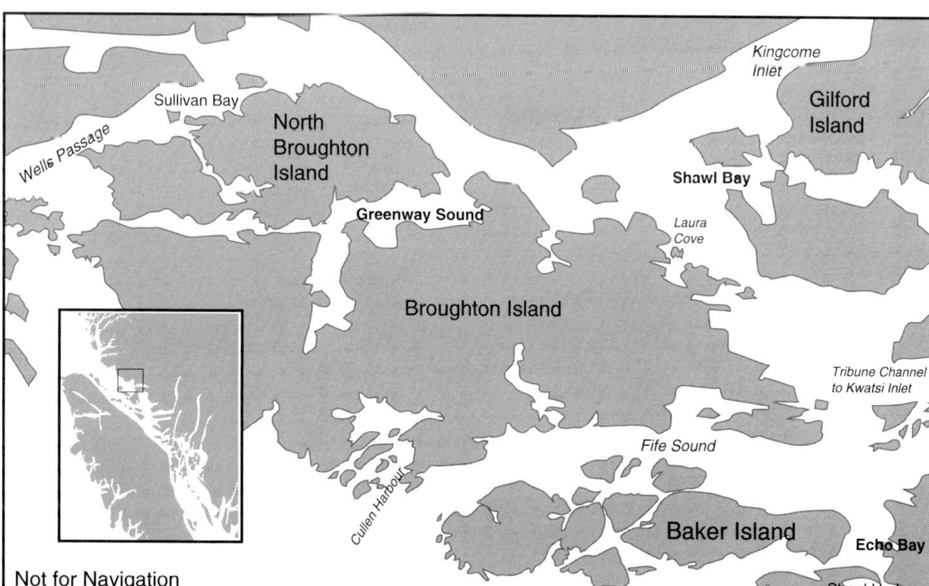

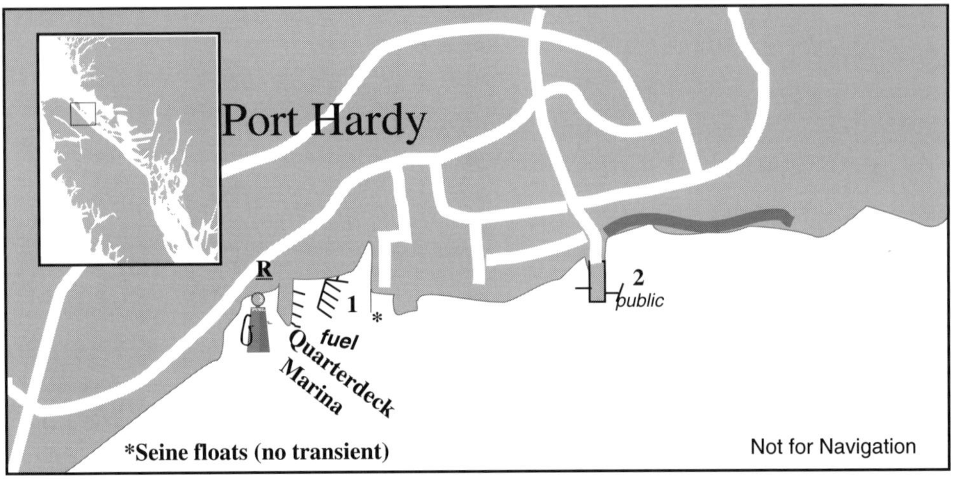

Quarterdeck Marina and Motor Inn

Box 910, 6555, Hardy Bay Road,
Port Hardy, BC. V0N 2P0
Phone: (250) 949-6551 Fax 949-7777
Chart 3548, 3605 VHF 73
email: *quarterdk@capescott.net*

Marina:

Moorage to 150 feet. 40 room Motor Inn. Reservations required from June 15th on. **Power** 15, 30 amp. **Water.**
60 ton travel lift. Full repairs available. Pressure wash. Marine store, Charts, ice, dry ice, propane,natural gas, fishing tackle.

Services:

Fuel. Showers, laundry. **Launch ramp**. Flights, bus service, ferry nearby. Pub restaurant ashore. Taxi and limo service to airport, BC Ferries and shopping.

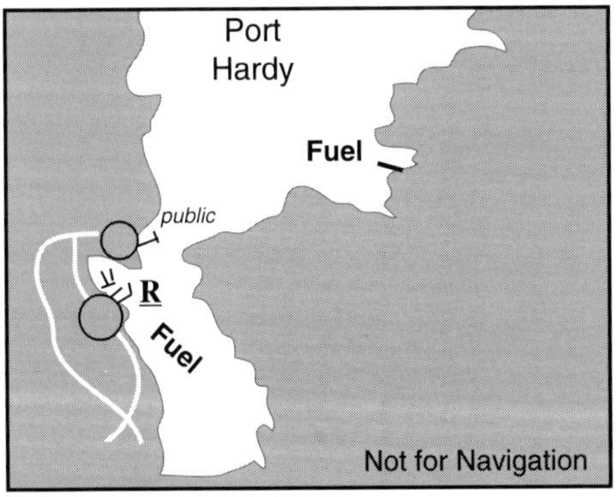

1. Port Hardy (Public)

(Inside breakwater–Fisherman's Wharf)
Port Hardy Harbour Authority,
P.O. Box 68, Port Hardy, V0N 2P0.
Phone (250) 949-6332 Fax 949-7433
Chart 3548, 3605
Manager - Rick Davidge. Also Seagate.
• Float length 574 m
Launch ramp • Breakwater • Grid • Garbage • Waste oil disposal • **Water** • Lights • **Power**–15, 20 amp • Public phone • **Washrooms** • Customs • **Fuel** nearby • Near city restaurants, services, shops. All services and facilities in Port Hardy and local marina. Ferries nearby for northern route.

2. Port Hardy (Public)

Transport Canada–Seagate Pier
Phone (250) 949-6332
Chart 3548, 3605
• Float length 250 m • **Garbage** • Waste oil disposal • **Water** • **Power** • **Moorage**–Commercial but some transient when space available. Floats in place May to September only. Excellent scuba diving nearby. Enquire at dive facilities in town. Nearby departure for Ferries north. Coast Guard station adjacent. Downtown Port Hardy location with access to shops and services.

Hurst Island

God's Pocket

Bill Weeks & Annie Geschi
P.O. Box 471, Port Hardy,
BC. V0N 2P0
Phone: (250) 949-9221
Chart 3549, 3605 VHF 73

Marina:

Moorage: Limited, sheltered docks. Washrooms, laundry. Some accommodation on availability.

Services:

Cabins, excellent meals. Scuba diving. Kayaking.
Some provisions, gifts, arts and crafts.
Coffee shop, lunches, pies.

Entertainment:

Walking trails on island. Hiking.

Bull Harbour

Hope Island
Fisheries & Oceans dock
Refer to chart 3549
Float length 35 m
Anchor in bay. Use dinghy.

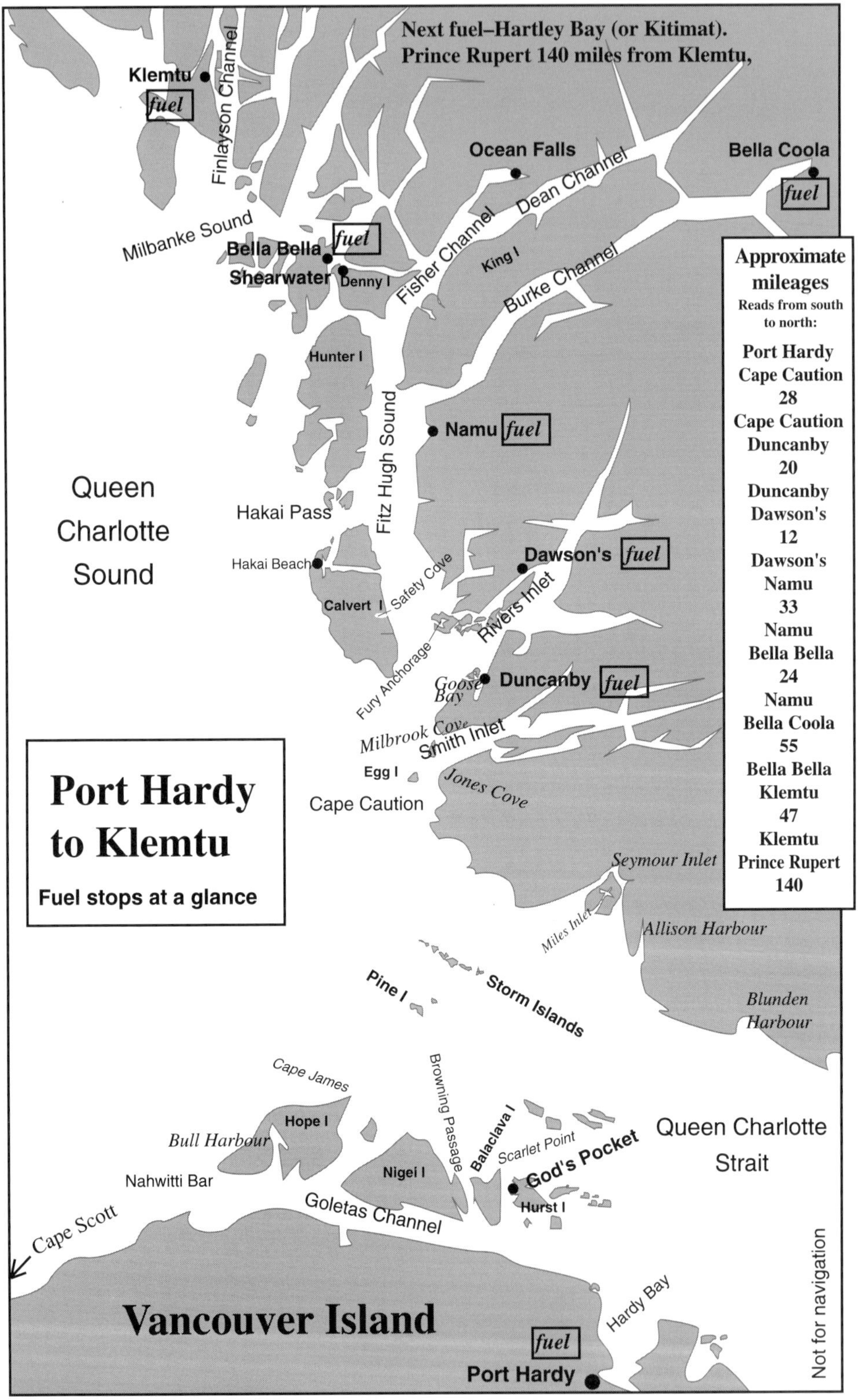

Next fuel–Hartley Bay (or Kitimat).
Prince Rupert 140 miles from Klemtu,
Klemtu
fuel
Finlayson Channel
Ocean Falls
Bella Coola
fuel
Dean Channel
Milbanke Sound
Bella Bella
fuel
Shearwater
Denny I
Fisher Channel
King I
Burke Channel
Hunter I
Fitz Hugh Sound
Namu
fuel
Queen
Charlotte
Sound
Hakai Pass
Hakai Beach
Dawson's
fuel
Safety Cove
Calvert I
Rivers Inlet
Fury Anchorage
Goose Bay
Duncanby
fuel
Milbrook Cove
Smith Inlet
Egg I
Cape Caution
Jones Cove
Port Hardy
to Klemtu
Fuel stops at a glance
Approximate mileages
Reads from south to north:
Port Hardy
Cape Caution
28
Cape Caution
Duncanby
20
Duncanby
Dawson's
12
Dawson's
Namu
33
Namu
Bella Bella
24
Namu
Bella Coola
55
Bella Bella
Klemtu
47
Klemtu
Prince Rupert
140
Seymour Inlet
Miles Inlet
Allison Harbour
Pine I
Storm Islands
Blunden Harbour
Cape James
Browning Passage
Hope I
Bull Harbour
Balaclava I
Scarlet Point
God's Pocket
Queen Charlotte
Strait
Nigei I
Nahwitti Bar
Goletas Channel
Hurst I
Cape Scott
Vancouver Island
Hardy Bay
fuel
Port Hardy
Not for navigation

The North Coast

Beyond Cape Caution

To reach Rivers Inlet from Port Hardy it is necessary to round Cape Caution and pass the entrance to Smith Inlet. There are no facilities between God's Pocket just out of Port Hardy, and Rivers Inlet. If weather is a deterrent temporary anchorage is possible off Hurst Island or moorage at God's Pocket dock if space is available. Bull Harbour, slightly out of the way for a passage around Cape Caution is a good anchorage. It is shallow off Cape Caution and many yachtsmen round the Cape about five miles off. Local mariners cut close to the Cape and take passage behind the islands, rocks and islets off Smith Inlet to slide around into Rivers Inlet off Goose Bay. Choose to round Cape Caution in gentle wind and sea conditions and be wary of fog. On the passage around Cape Caution one can find temporary and some good protected overnight anchorages at places such as Blunden Harbour, Allison Harbour, Miles Inlet, Seymour Inlet, Jones Cove and Milbrook Cove. Please refer to ***Anchorages and Marine Parks***.

The first fuel stop is at Goose Bay–Duncanby Landing or farther up Rivers Inlet at Dawson's Landing. Some mariners travel past Rivers Inlet, stopping, if necessary for overnight anchorage, at Safety Cove on Calvert Island, or at Fury Anchorage at the entrance to Rivers Inlet.

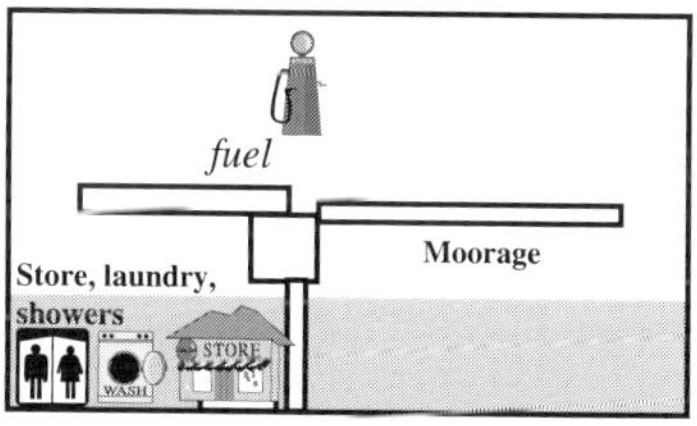

Rivers Inlet

Duncanby Store & Marina

Ken Gillis
Rivers Inlet P.O. BC. V0N 1M0.
Phone: (250) 949-2101
Chart 3934 VHF 6

Located at Goose Bay, Rivers Inlet.

Marina services:

Moorage. Water, power–15 amp.
Fuel: gas diesel, propane.

Facilities:

Showers, laundry, washrooms.
Store: charts, books, tackle, bait, fishing licences, groceries. Ice subject to availability. Liquor store.

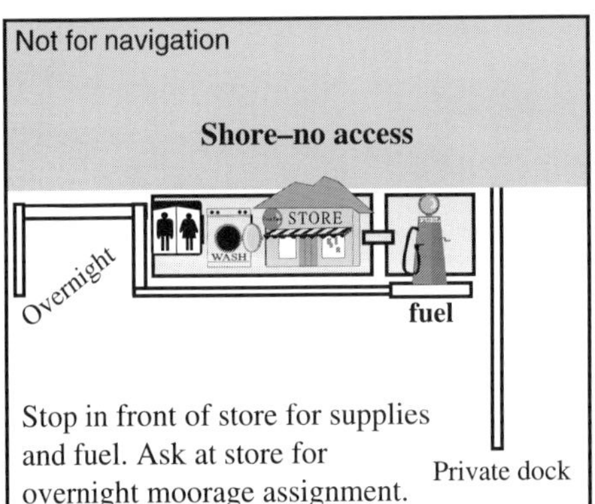

Tied up to the fuel and store dock at Dawson's Landing. Inside the store there is always a warm welcome. Opposite page: The dock at Hakai Pass Resort looking down Pruth Harbour.

Dawson's Landing & General Store

Robert and Nola Bachen
Dawson's Landing, BC. V0N 1M0
Phone: (250) 949-3533 Fax 949-2111
Chart 3934, 3932 VHF 6, 85

Marina services:

Moorage. Water. Fuel: gas, diesel, oils.

Customer Services:

Showers, laundry.
General store: tackle, fishing licences, bait, groceries, liquor agency.
Post office, gifts, toys, souvenirs, books, charts, supplies for marine maintenance and repairs.

Note: No power at docks.
Excellent anchorages in nearby islands.

Hakai Pass

Hakai Beach Resort

Dock manager

Pruth Harbour, Lewall Inlet

P.O. Box 3819, Smithers, BC. V0J 2N0

Phone (250) 847-9300

Charts 3784, 3727.

Anchor in bay. This is not a public dock or marina. Shore access–tie up small dinghies under ramp inside marina.

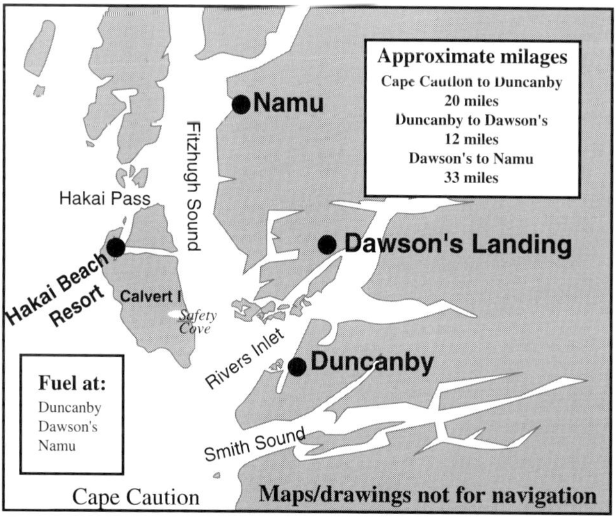

Resort offers restaurant and accommodations when available. Store–gifts, film, snacks, cappuccino. Walk to white, sandy beach on open Pacific.

Anchorage–Pruth Bay.

Boats at anchor receive welcome flyer.

Guided fishing excursions available.

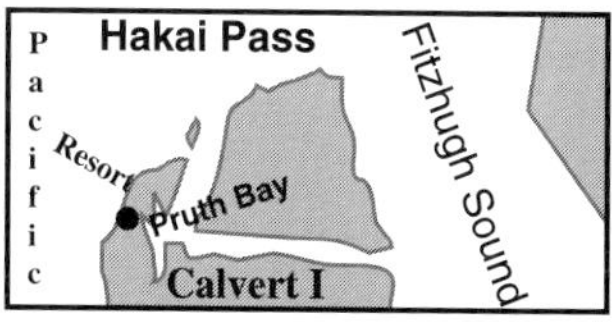

Namu to
Bella Coola
55 miles
Bella Bella to
Ocean Falls
30 miles
Bella Bella
to Klemtu
47 miles

Namu

Namu

New management
Namu, BC. V0N 1M0
VHF 6
Charts 3785, 3727, 3784

Marina services:

Moorage. No power at docks.
Water at main dock.
Fuel: gas, diesel, oils.
Customer services: (Subject to changes) Historically a store offering tackle, fishing licences, bait, ice, groceries, liquor, post office, gifts, charts, books and marine supplies has been operational. There was also a ways and marine repairs were available. Toperty had a restaurant and it is expected to be open in 1999.
This is a popular area for kayaking.
Activities: Hiking trails. Trout fishing in Namu Lake.
This stop is an interesting historic and archaeological site.

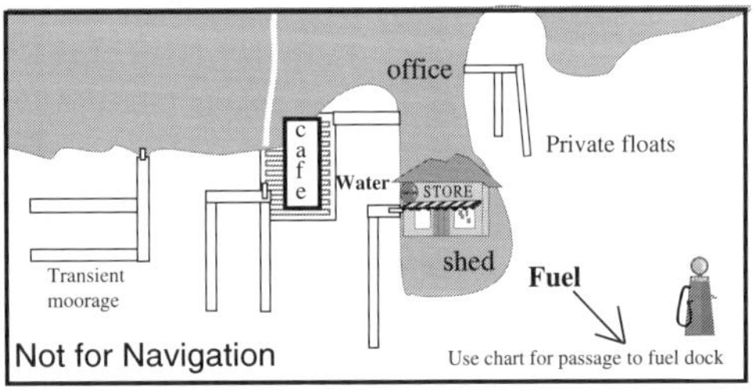

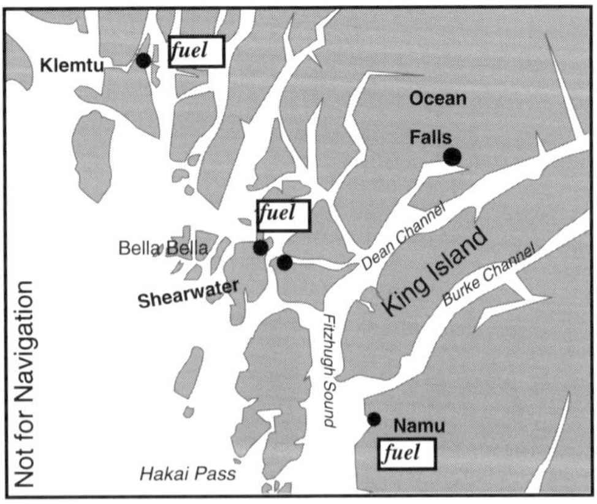

Shearwater

Shearwater Marine

Al Tite
P.O. Box 68, Shearwater,
Bella Bella, BC. V0T 1B0
Phone: (250) 957-2305 Fax 957-2422
Vancouver (604) 270-6204 Fax 270-4974
Charts 3720, 3785, 3787
Web: *www.shearwater.ca*

Marina services:

Moorage. 2000 feet of docks (1,000 new concrete dock) **Water, Power** 15, 30 amps **Laundry, showers, washrooms.** Propane. Public phones. 100 ton haul outs. Marine repairs. travel lift. Boat launch.

Customer services:

Store: tackle, fishing licences, bait, groceries, gifts. Charts, marine supplies. Pub. Off-sales.
Restaurant. (250) 957-2366
40 room hotel. 3,000 foot airstrip nearby. Fishing charters, kayaking, boat rentals. Camping. Scuba diving and air station. Daily floatplane service nearby.
B.C. Ferry terminal.

Bella Bella

Bella Bella

Bella Bella, BC. V0T 1B0
(250) 957-2440
Chart 3720, 3785, 3787

Marina:

Fuel: gas, diesel, stove oil, lubricants.

Customer services:

Moorage. Water, public phone.
Store. Hotel with public phones, **showers,** pub, coffee shop. BC ferry stops weekly. RCMP station. Hospital. Liquor. Repairs, haul out facility. Short distance to Shearwater Marine Group. Bella Bella is a native village.

Ocean Falls

Namu to Bella Coola 55 miles

Namu to Bella Bella 24 miles

Bella Bella to Ocean Falls 30 miles

Bella Bella to Klemtu 47 miles

Ocean Falls to Bella Coola 50 miles

Namu to Kitimat 181 miles

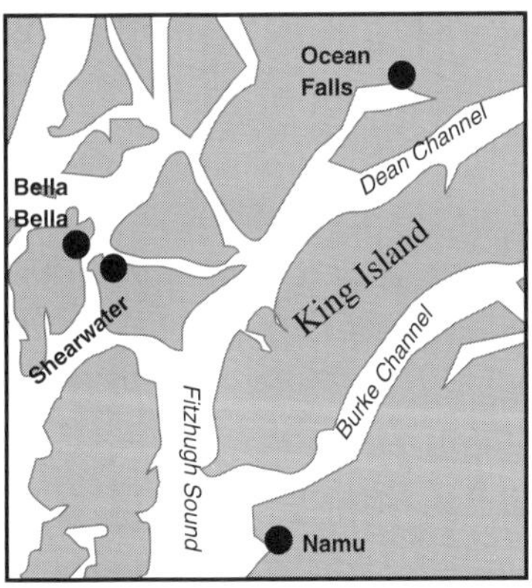

Docks at Bella Bella

Ocean Falls

Ocean Falls

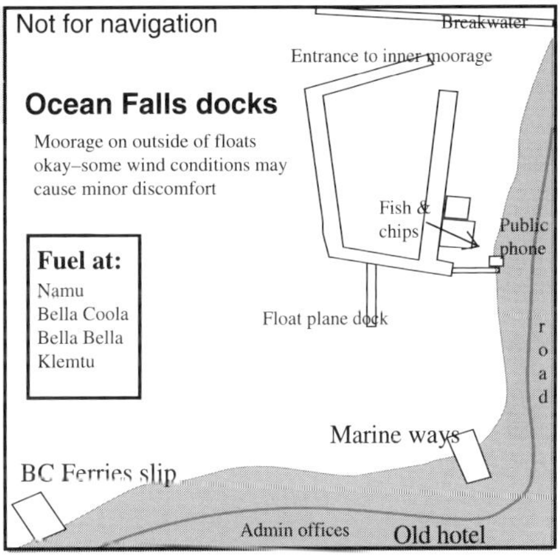

Top and bottom opposite page: Ocean Falls docks. Above: An old house at Ocean Falls. Below: The floats include a small fish and chips/ restaurant which is open several days a week.

Ocean Falls

Herb Carpenter
(Ocean Falls Yacht Club)
Ocean Falls, BC. V0T 1P0
Chart 3781, 3720, 3729.

Marina:

Moorage. Power 20 amps. **Water,** Fish cleaning station. Floating fish and chips restaurant on docks. Delivery service groceries to your boat. (250) 289-3333. Also fishing tackle, licences, bait, supplies. Phone at head of dock. Post office and town administration offices in old courthouse.
Marine ways, repairs available. Powerboats to about 45 feet.
No cellular reception.
Walking old roads and roadway to lake.
Old powerhouse and mill barricaded for safety.

Bella Coola main street.

Bella Coola

Bella Coola

Kevin O'Neill
Bella Coola, BC.
V0T 1B0
Charts 3730, 3729

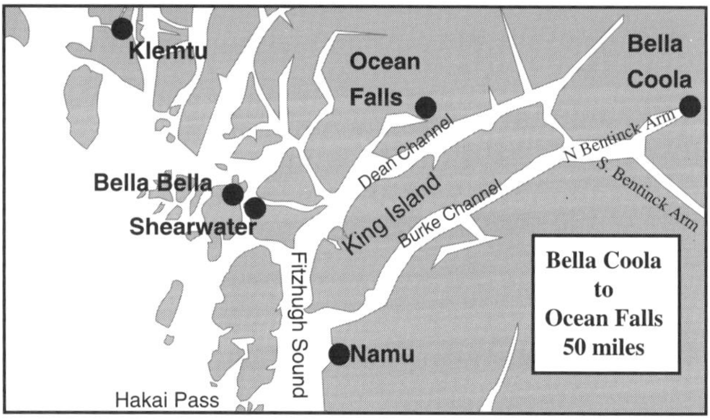

Marina:

Fuel, (closed week-ends). gas, diesel. **Power** 20 amps. **Water.** Garbage disposal. On site **showers, washrooms**, public phones. One and a half miles to uptown **laundry**, showers, washrooms. Stores, restaurants, co-op, museum, motels, taxi, car rentals. Tackle, fishing licences, bait, groceries, liquor, gifts. Charts, marine supplies.
Cedar Inn Restaurant. Kopas Store.
No food services at marina.
Rafting necessary at pleasure boat moorage near shore on east side of marina.
No VHF radio reception. Wharfinger available 9-11am only (weekdays).

If you ask the townsfolk at Bella Coola where their favourite hot springs is located, they may tell you. Just down North Bentinck Arm to South Bentinck Arm and turn south. Several miles down the arm just beyond Bensins Island and on the opposite shore is Talheo hot springs at the mouth of Hotsprings Creek. Anchor off and row ashore for a hot soak in the sulphur waters of this delightful pool. It has been improved by the local people of Bella Coola so please respect it and leave it clean.

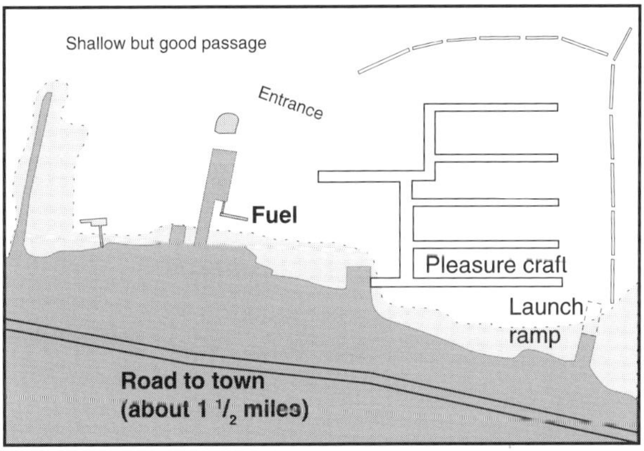

Above, centre and top: Views of Bella Coola harbour and dock plan.

Bella Coola may be reached overland but it is far easier by water. And the town is worth a visit. The walk into town from the harbour is pleasant, about one and a half miles, with magnificent views over the delta and the high adjacent mountains. History was made in this area with the arrival of Alexander McKenzie in July, 1793 when he completed the first recorded crossing of North America–even before Lewis and Clark, by some 13 years. He was not far behind Captain Vancouver who had passed through the area only seven weeks previously.

In Bella Coola visit the museum where the curator may have some surprise information in store for you. She is Rene, daughter of Cliff Kopas, early settler and author of the book *Bella Coola* as well as founder of Kopas General Store.

In summer months the town is busy with festivities ranging from an indigenous canoe festival, to a rodeo, theatre, music festival and a fall fair and parade.

Klemtu

Klemtu Marine Station

Steve Robinson
Klemtu, BC. V0T 1L0
Phone (250) 839-1255
Charts 3711, 3734, 3902 **(3728 enroute)**
CB Call sign 44

Marina:

Fuel: gas, diesel, stove oil. Ice. **Water.** Store at fuel dock open 6 days a week: Mon–Sat 8.30 to 12 noon. 1–5pm. Coffee shop. Public phone.
Moorage nearby at public dock. Up town store: Propane. Groceries—open 9am to 12 noon. 1-5pm. 6.30-10pm.

Klemtu is the last significant stop before Prince Rupert or Kitimat.
Butedale has some fuel (reported summer 1998) but don't rely on it as a stop unless you check first.
Fuel at Hartley Bay–call to confirm.
Distance Klemtu to Prince Rupert 150 miles.

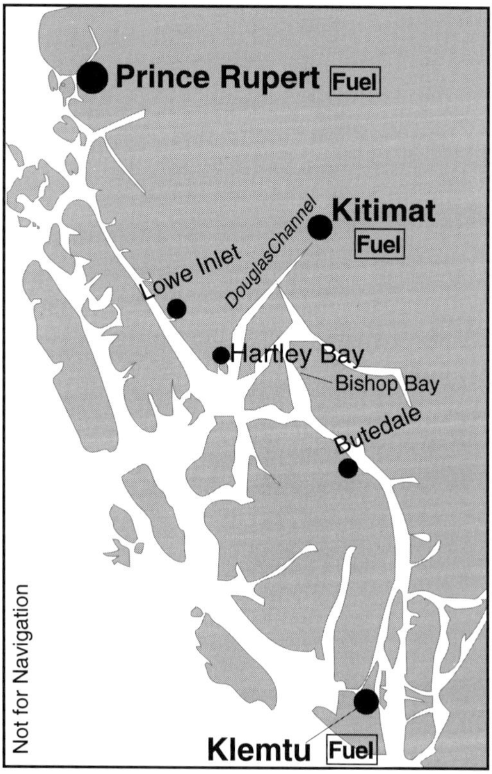

Hartley Bay

Hartley Bay (Public)

Ed Robinson
Hartley Bay, BC. V0T 1L0
BC Tel 60. Chart 3711, 3742
Phone (250) 841-2500
Marina: Moorage at public docks behind breakwater.
Fuel. Power 20 amps**. Water**. Groceries. Post office. General store. Emergency helipad. Nearby mooring buoys in Stewart Channel. Smaller gas boats top up at Hartley Bay during the 150 mile trip from Klemtu to Prince Rupert.

Church and marina (top) at Hartley Bay

Kitimat

M.K. Bay Marina

P.O. Box 220, Kitimat, BC. V2C 2G7.
Phone (250) 632-6401
Fax: (250) 632-6889 VHF 68
email: *wdevins@sno.net*
Chart 3743, 3736 Launch ramp
Marina: Fuel: gas, diesel, oils. **Water. Power** 15, 30 amp. **Laundry, showers, washrooms.** Public phone. Repairs. Haulouts to 25 tons.
Facilities: 11 kilometers to town. Taxi.

Moon Bay Marina

Peter Jensen
P.O. Box 196, Kitimat, BC. V8C 2G7
Phone (250) 632-4655. Monitors Channel 16–Call vessel *Moon Bay*.
Fuel. Water. Power available. Marine repairs and service. Rental fishing boats. Owner Don Pearson operates a tug with supplies and some **fuel** in Douglas Channel/Bishop Bay area in summer. The tug is a base station for small boats wanting to join group excursions: it offers showers and other amenities at sea.
Up town Kitimat 10 kilometers : Taxi. Stores, liquor**, post office**, **hospital.**

Prince Rupert

Prince Rupert, BC is a fishing town. Its docks in season are brimming with commercial vessels rafted as many as six deep at the town's two major public docks. Moorage overnight for pleasure craft is best at the Prince Rupert Yacht Club where the spaces are tight and subject to being available only if members are away temporarily. It is necessary to call ahead to reserve space, although frequently boats arriving early in the day will be allocated a slip immediately or later in the day.

Directly above the marina there are friendly services for the transient mariner, from laundry and showers to nearby restaurants, pubs, shopping and marine repairs. Arts and craft of local people can be found in stores up town. A good hotel with restaurant and pub sit atop the cliff overlooking the harbour.

Fuel is available at a number of fuel docks along the waterfront. Prince Rupert is the final major stop before continuing to Alaska. Mariners learn while talking to other yachtsmen during stops along the British Columbia coast that entering Alaska may not always be straight forward. Wind can prevent vessels, especially slow cruising yachts, from making the Rupert to Ketchikan run all in one stretch. And the reason most skippers plan it in one step is the belief that one has to do so as a customs clearing requirement. What one learns is that it is possible to stop en route, either on the Canadian side at Dundas Island or on the Alaskan side several miles up channel from Cape Fox at Foggy Bay. If there is any doubt whatever that you can make the run in one leg, it is best to call Alaska customs and ask for permission to stop at a specific place en route. Remember also that Alaska, unlike BC, remains on Standard Time all year.

Once in Ketchikan there is much to do and see, from a visit to the Totem Heritage Center to walking the old town with its infamous Creek Street, once the place of bawdy houses and frontier life at its most colourful.

It's a tourist town with several large cruise ships at a time lining the docks right alongside the downtown main streets. And it's a town which will prompt you to stay a while, test the delights of fine cuisine and then go on deeper into the Alaskan South West or Panhandle, whichever you prefer to call it. There are numerous books and guides on cruising Alaska. Once in the State you will find easy access to lots of cruising options with the easy availability of fuel and services along the way.

The Prince Rupert Yacht Club, above, is located in the distance on the east shore of the harbour, opposite.

Prince Rupert Yacht Club

121 George Hills Way
Prince Rupert, BC. V8J 1A3
Phone (250) 624-4317
Chart 3957, 3958, 3955 VHF 72

Marina services:

Moorage. Power 15 & 20 amps. **Water. Shower,** ice, bait, tidal grid, power wash available. No reciprocal club moorage. All visitors pay moorage fee towards upkeep.

Adjacent facilities:

Fuel, Esso (628-6277), Petro Canada (624-4106), Chevron (624-3316)–next to PRYC–Gas, diesel, stove oil, water, **showers, laundry,** ice. Snacks.
Up town Prince Rupert is five minutes walk. Also restaurants (Smiles, Breakers Pub, Crest Hotel).
Note: Five miles an hour speed limit strictly enforced in Prince Rupert Harbour. Boats are monitored by radar and any one caught exceeding the limit will be fined $300 by ticket in the mail.
Anchorages at Pilsbury Cove and Tuck Inlet.
Public docks nearby–busy during fishing season: rafting. Fishing openings usually begin 6pm Sundays.

Public docks:

Rushbrooke Harbour

Robert Small
P.O. Box 1026, Prince Rupert, BC. V8J 4B7. Chart 3958 VHF 10
Phone (250) 624-9400

Marina services:

Large marina with about 400 slips. Power 15 amps. **Water,** portable toilet, public phone, free launch ramp adjacent. Garbage disposal. 20 minutes walk into town.

Cow Bay

Robert Small
P.O. Box 1026, Prince Rupert, BC. V8J 4B7. Chart 3958 VHF 10
Phone (250) 624-9400
(minimal space for brief stops)

Fairview Harbour

Robert Small
P.O. Box 1026, Prince Rupert, BC. V8J 4B7. VHF 10 Chart 3958
Phone (250) 627-3127
Marina Services: Fishing harbour with limited moorage when available.

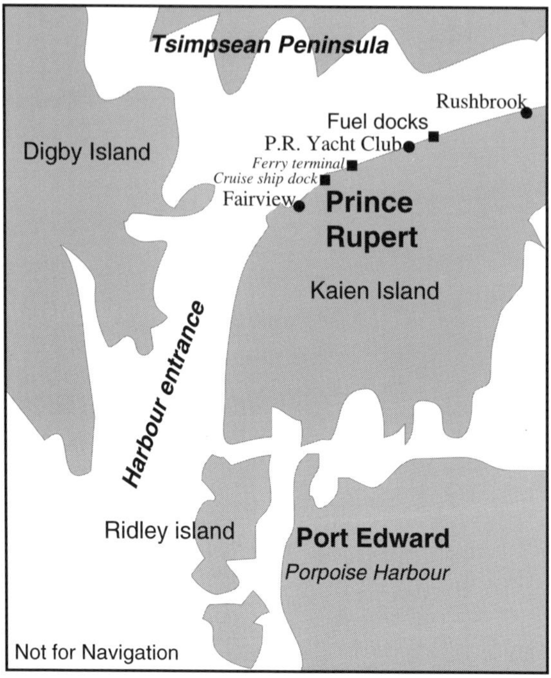

Opposite Prince Rupert Yacht Club. Call on channel 72 for moorage reservations.

Porpoise Harbour Marina

Rick Hill
email: *peharbor@citytel.net*
P.O. Box 1820, Port Edward, BC. V0V 1G0
Phone (250) 628-3211
Chart 3955, 3858 VHF 79A

Marina services: Fishing harbour–public moorage. **Fuel. Power** 20 amps. **Water, laundry, washrooms,** public phone. Travel lift.

Adjacent facilities: Cab service to Prince Rupert–Skeena Taxi 624-2185. Convenience store. North Pacific Museum.

The Queen Charlottes

For those crossing to the Queen Charlottes there are a few marinas available for transient moorage. There is a large marina at Sandspit and several public marinas in other parts of the islands as listed below and on the following page.

Sandspit Harbour Marina

email:*harbour@sandspitqci.com*
www.sandspitqci.com/harbour
P.O. Box 477, Sandspit, BC. V0T 1T0
Phone (250) 637-5700 fax 637-5369
Chart 3890, 3894 VHF 72A

Marina services: Moorage up to 100'. Power. Water, **showers, washrooms,** public phone. Launch ramp.

Adjacent facilities: Car rentals, shops and city services and amenities.

QC Small Craft Harbour

Bob Olson
P.O. Box 68, Queen Charlotte V0T 1S0 BC. V0V 1G0
Phone (250) 559-4650
Chart 3890,3894
Marina services:
Public dock moorage
Water, garbage, **power,**

Port Clements Small Craft Harbour

Chris Marrs
P.O. Box 198, Port Clements V0T 1R0 BC.
Phone (250) 557-4295
fax (250) 557-4574
email:
portclem@qcislands.net
Chart 3893
caution: shallow at low tide.

Delkatla Slough

Dale Otto
P.O. Box 35, Masset, BC V0T 1M0
Phone (250) 626-5487
fax 626-5193
Chart 3892,3895
VHF 06
Marina services: Fuel. Power 20, 30 amp. Water. Tidal grid.
Shower, laundry. Shops

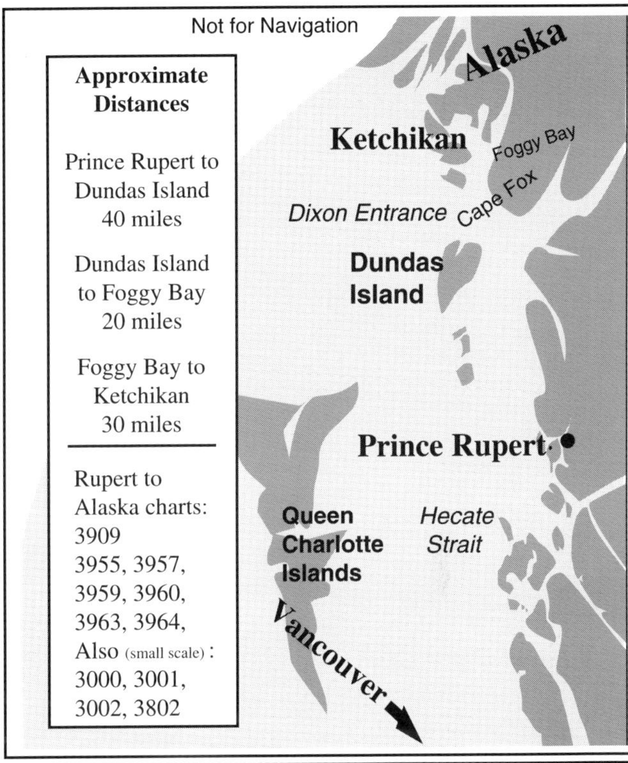

Next stop—Alaska:

If you are proceeding to Alaska, your fist stop will be Ketchikan, probably via a short stop at Foggy Bay. This protected bay is a short distance beyond Cape Fox which you pass as you venture across Dixon Entrance into Alaskan waters. If you are planning to stop in American territory before reaching Ketchikan, official customs port of entry, you must obtain permission from customs in Ketchikan first. This can be done by calling (907) 225-2254 and advising customs that you will need to stop at Foggy Bay. Ask for permission to do so rather than telling the customs officer that you are going to stop there. And if the weather permits a straight, uninterrupted run into Ketchikan simply dock at a convenient slip and call customs. They may ask you to walk up town to their offices to clear. Note: It is important to read the customs document you are issued, as it instructs you to notify customs at certain other Alaska points as you proceed farther along the coast. This is the case, even though you have cleared at Ketchikan (or elsewhere in Alaska). Customs hours are office hours, week-days only. Special arrangements are required for clearing at weekends.

A safe harbour to spend time waiting for suitable wind and sea conditions to cross Dixon Entrance is at Goose Bay on the north end of Dundas Island. If you set out from Prince Rupert in calm seas but the wind picks up and dictates a rough passage beyond Dundas stop a while and do some fishing, reading and relaxing. Then travel when conditions improve. At Dundas you are about 20 miles from Foggy Bay or about 70 from Ketchikan. Rupert to Ketchikan is 90 miles.

Above: Late afternoon in Tribune Channel looking west from the entrance of Kwatsi Inlet where a new marina is springing up adding to the popular nearby destinations at Greenway Sound, Sullivan and Echo Bay.

Inset: Infamous Creek Street, Ketchikan. Above: City docks in downtown Ketchikan. First stop en route to Alaska cruising destinations.

South Bound
on the West Coast of
Vancouver Island

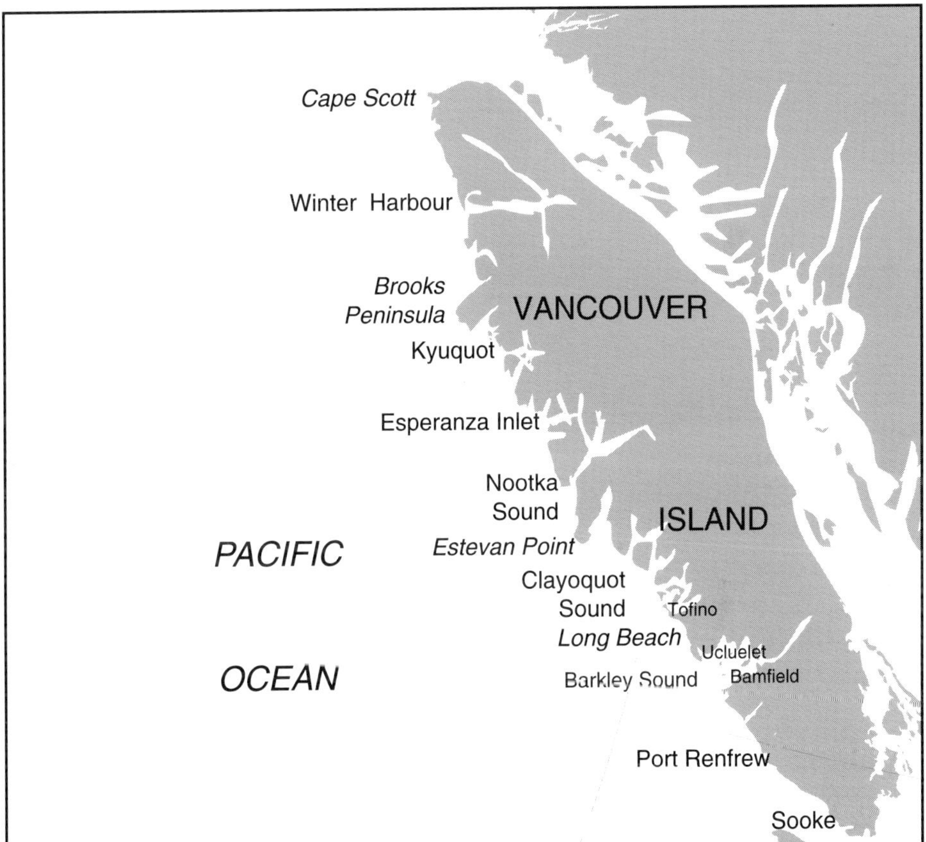

The west coast of Vancouver Island enjoys a short summer season of recreational boating. All year around, however, there is maritime activity catered to by a small number of marine facilities. These are scattered among the inlets and coves that indent the rugged coast and serve as a home away from home to those out working the coast. They serve also as havens of safety and replenishment for those seeking to extend their pleasure boating experiences in British Columbia. It is possible to travel along the Pacific coast of Vancouver Island yet remain in sheltered waters a good deal of the time. The open coast is broken up by numerous islands, large and small, forming inlets and protected waterways. One can travel along these waterways, poking out into the open Pacific during calm conditions and moving along to the next inlet which in some cases is no more than about 25 miles distant. The longest stretches of open water are between Juan de Fuca Strait and Barkley Sound in the south and between Quatsino Sound and Bull Harbour to the north.

Winter Harbour

Winter Harbour

Fisheries & Oceans dock
Charts 3686, 3617
Manager • Float length 171 metres •
96 m wharf • Garbage disposal • Derrick • Water •
Plus: Showers, (Laundry for commercial boats—
pleasure boat owners should ask permission).
Public phone ashore. General store nearby.

The photograph above was taken some years ago at Winter Harbour. Wildlife presence and the "main road" walkway have not changed.

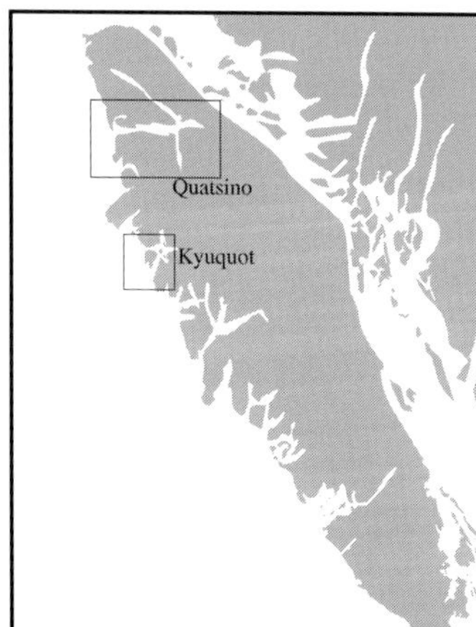

BC Packers
Winter Harbour, BC.

Services:

Fuel–gas, diesel.

Quatsino

Transport Canada dock
Charts 3681, 3617
Manager • Float length
31 metres •
30 m wharf • Aircraft
float • Sheds

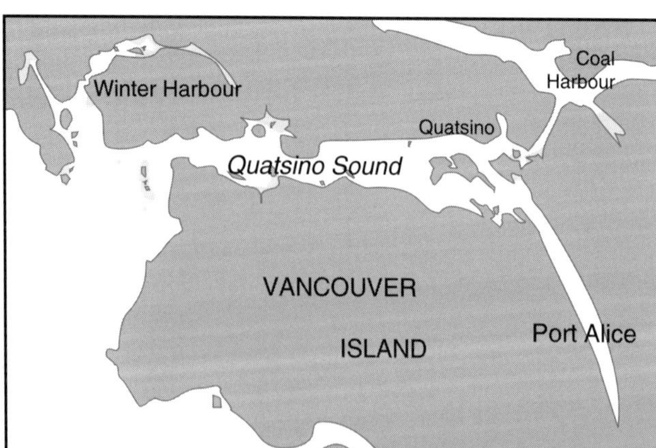

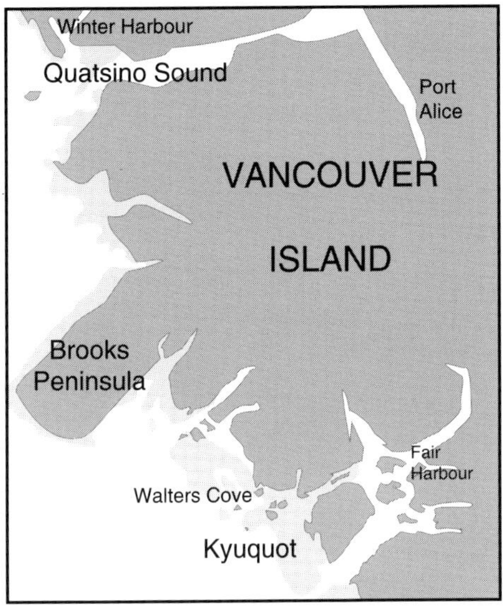

Quatsino Boatyards and Marina

Chris and Debbie Bradley
General Delivery, Quatsino BC.
V0N 2V0
Phone (250) 949-6651
Charts 3681, 3617 VHF 10

Marina services:

Moorage–summer transient available at times. **Power** 110/220.
Boatyard haulouts and repairs.

Nearby:

Walk the roads around Quatsino. The Narrows just beyond this point leads on to Coal Harbour, an historic whaling station. Down Neroutsos Inlet you will find Port Alice, where the only possibility of a casual stop is at the yacht club.

Port Alice

Yacht Club
Charts 3681, 3617
Temporary mooring (unofficial) at club docks.

Coal Harbour

Anchor Petroleum
VHF 10
Charts 3681, 3617 VHF 10

Marina services:

Moorage–no moorage. **Fuel** dock for commercial trade but pleasure craft are welcome.

Coal Harbour.
Transport Canada dock
Charts 3681, 3617
Float length 94 metres • Garbage/ waste oil disposal • Derrick • Lights • Power •

Kyuquot

Walters Cove

Susan Bostrom
P.O. Kyuquot, BC.
Chart 3651, 3682, 3683

Marina services:

Moorage–public dock. Post office and store at head of dock. Cafe adjacent to post office.

Kyuquot (Houpsitas)
Fisheries & Oceans dock
Charts 3651, 3682, 3683
Float length 82 m •

Kyuquot Transport Canada dock
Charts 3651, 3682, 3683
Float length 103 metres •
30 m wharf • Derrick • Sheds

Fair Harbour
Transport Canada dock
Charts 3682, 3623
Manager • Float length 28 metres •
24 m wharf • Derrick • Sheds
Launch Ramp adjacent.

Tahsis

Westview Marina

Mike and Cathy Daynes
P.O. Box 481, Tahsis, BC. V0P 1X0
Phone (250) 934-7622 Fax: 934-6445
Chart 3663, 3665 VHF 73 06
Moorage–summer transient available at times. Reservations recommended.
Fuel-gas, diesel. **Power** 15, 30 amp.
Water. Store. Quarter mile to Tahsis for **launch ramp**. Also for **laundry, showers,** pool, restaurants, supermarket, hospital, RCMP, Info Centre and Museum.
Beware of logs in harbour as you approach Tahsis and the marina.

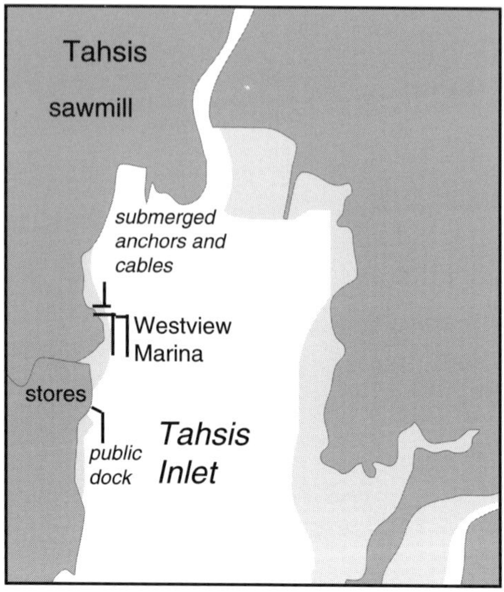

Tahsis

Transport Canada dock
Charts 3665, 3663
Manager • Float length 30 metres • 42 m wharf • Garbage disposal • Derrick • Lights • Sheds • Public phone ashore • General store • liquor agency • Post Office • shops and accommodations nearby.

Esperanza

Box 398, Tahsis, BC. V0P 1X0
Phone (250) 949-6651
Chart 3663, 3662 VHF 06, 17

Transport Canada (Hospital Floats)
Manager • Float length 42 metres • Water • Lights • **Moorage**–summer transient space available.

Esperanza Public dock/fuel dock.
Float length 15 m • Wharf 29 m • Derrick • Lights. **Fuel–**gas, diesel, stove oil, naptha, kerosene • **Water.** **Mini store • Showers. Laundry.** **Phone** ashore (emergency use only). Food available at Nootka Mission camp.

Zeballos

Weston Enterprises

Tom and Alice Weston
Box 100, Zeballos, BC. V0P 2A0
Phone (250) 761-4201
Fax (250) 761-4618
Chart 3663, 3664 VHF 68
Fuel gas, diesel, propane, ice.
Seaplane landing and fuel dock
Moorage at nearby public docks.

Zeballos Small Craft Harbour

Fisheries & Oceans dock
A. Coburn
P.O. Box 99, Zeballos, BC. V0P 2A0
Phone/fax (250) 761-4333
Chart 3663, 3604. VHF 68
Float length 191 metres • 96 m wharf • Garbage/oil disposal • Water • Lights • **Power** 15 amp. Fuel adjacent (Weston)
Launch ramp—all tides. Available free.
Transport Canada dock nearby.

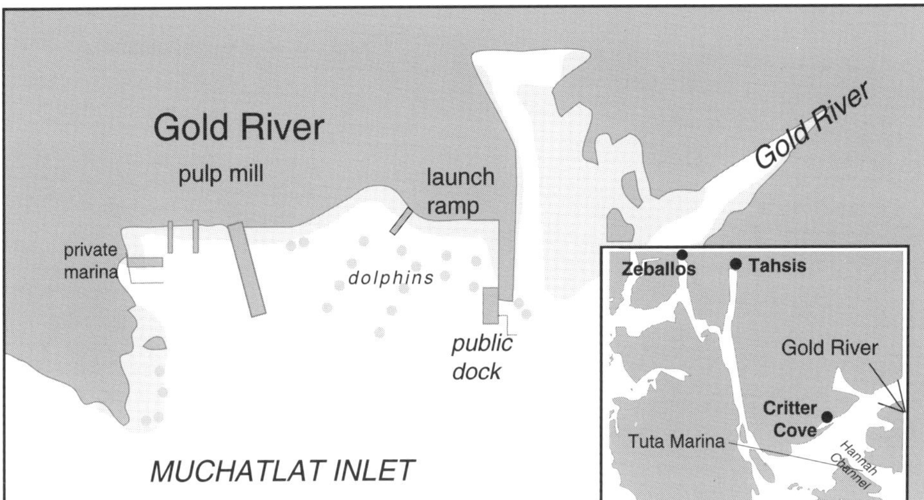

The town of Gold River is about eight miles inland. The harbour is subject to wind and tidal conditions making it not suitable for moorage. A ramp adjacent to the river breakwater serves small boat launching. Overnight at Tuta Marina or Critter Cove Marina.

Gold River

Critter Cove Marina

Cameron and Dean Forbes
Box 1118, Gold River, BC. V0P 1G0
Phone (250) 283-7364 (604) 886-7667
Chart 3665, 3664,3681 VHF 07
Moorage on 1000 feet of dock space as available in summer May 16 to September 15. Call for reservations. Anchoring nearby. Rental cabins and suites available.
Fuel–gas. Store, licences, bait.
Coffee House. Known for their chowder and home-made pies.
Showers, washrooms, water.
Launch ramp at Gold River.

Tuta Marina

Larry and Shirley Andrews
Box 765, Gold River, BC. V0P 1G0
Phone (250) 283-7550
Chart 3664 Located in Hanna Channel, Nootka Sound. Open summer only Victoria Day to Labour Day
Fuel–gas, oil. Boat launch.
Store, licences, bait.
Showers, laundry.

Gold River

Transport Canada dock
Charts 3664, 3665
Manager • Float length 47 metres • wharf • Aircraft float • Garbage disposal • Water • Lights • Power • Telephone • Sheds • **Launch Ramp**

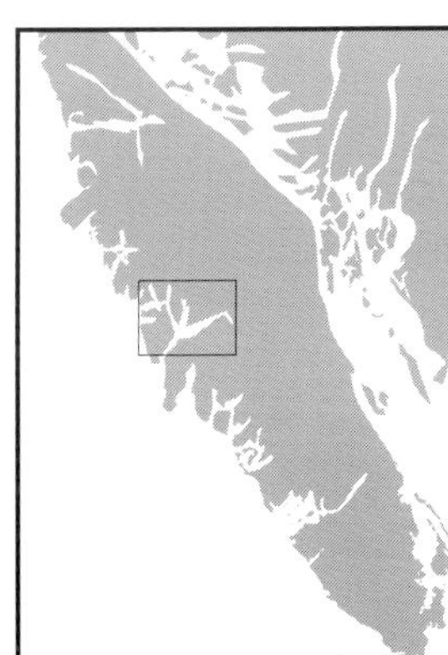

Fourth St. Docks at Tofino

Hot Springs Cove

public dock
Chart 3643, 3648, 3640
VHF 10 ***Marine Park***

Photograph above: Fourth Street pub dock at Tofino. Fuel is available at Method Marine (opposite). The service station, part of the marina complex, is a major fuel stop on the west coast.
Below: Weigh West Resort docks. This is a lodge that attracts longer term small boats. If there is space overnight moorage is provided. Larger boats check out the long floats on the outside. There is an adjacent restaurant and other facilities.

Ahousat Matilda Inlet

Marktosis–Fisheries & Oceans
Hugh Clarke
Charts 3643, 3648, 3640
Float length 232 metres • 21 m wharf • Derrick • Lights • Sheds • Public phone. **General store.** Cafe. Post Office. **Fuel.** Marine Ways. **Water.** Nearby is the Gibson Marine Park and Hot Springs. See ***Anchorages and Marine Parks.***

Tofino

Method Marine Supply

Steve Bernard
Box 219, 380 Main Street, Tofino. BC, V0R 2Z0.
Phone (250) 725-3251 Fax 725-2111
VHF 06 *(formerly Ocean West)*
Charts 3685, 3649, 3640, 3603.
Moorage–summer transient at adjacent marina docks. Reservations recommended in summer. **Power** 20 amp. **Fuel**: gas, diesel, stove oil. **Water**, ice, propane, **laundry, showers, toilets.**
Chandlery, marine service station–parts and accessories. Tackle shop–snacks, coffee. Scuba air fills. Charts, bait.
Nearby
Launch ramp, accommodations, art gallery, restaurant, liquor, bakery, supermarket and other stores.

Weigh West Marine Resort

Box 69, 385 Campbell St, Tofino.
Phone (250) 725-3277
Guest moorage, charters, marina and pub.

Tofino

Fourth Street, Tofino
Fisheries & Oceans dock
Charts 3685, 3649, 3640
Manager • Float length 444 metres • 15 m wharf • Breakwater • Grid • Garbage/oil disposal • Derrick • Water • Lights • Power • Sheds • **Launch ramp.** Public phone ashore. General stores, liquor agency, Post Office, shops, laundry and accommodations nearby.

Armitage Point, Tofino
Fisheries & Oceans dock
Charts 3685, 3649, 3640
Manager • Float length 121 metres • 30 m wharf • Garbage disposal • Derrick • Lights. Public phone ashore. General stores, liquor agency, Post Office, shops, laundry and accommodations in town, nearby.

Above: Method Marine docks. Note the native village of Opitsat across the harbour.
Inset: Armitage Point public docks.

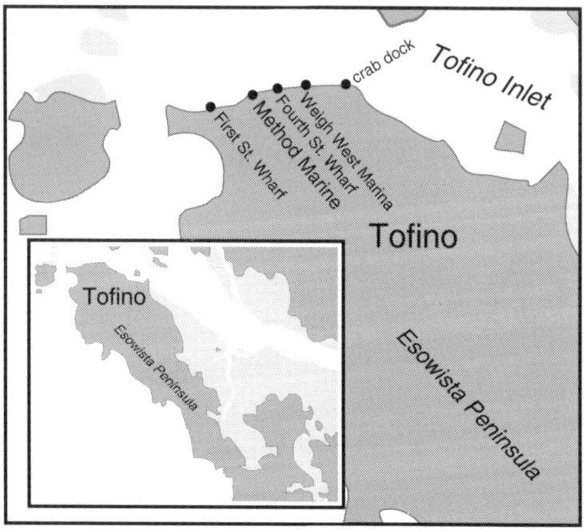

Ucluelet Boat Harbour docks–the entrance from shore in photo at top, entrance passage from the inlet in the photo above.
Diagrams show–Tofino on the Esowista Peninsula. Note the shallows throughout the harbour. Use a large scale chart for safe navigation. Ucluelet and Ucluth Peninsula. Note approximate location of primary marinas. Barkley Sound with Ucluelet and Bamfield locations.

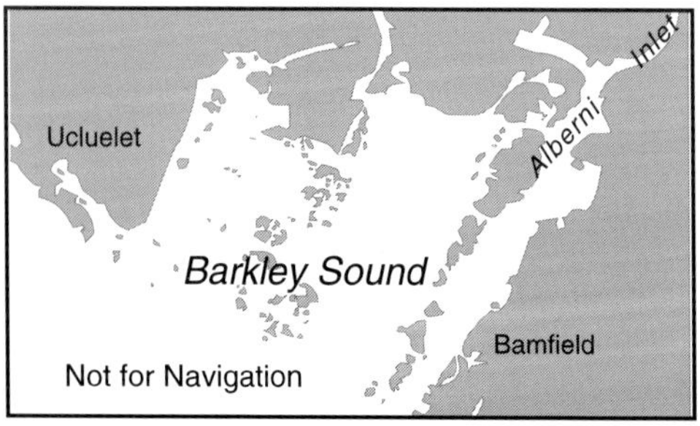

Ucluelet

Island West Resort

P.O. Box 32 Ucluelet, BC. V0R 3A0
Phone (250) 726-7515 Fax 726-4414
Chart 3646, 3671, 3603 VHF 69
email: *fish@islandwestresort.com*
Web: *www.islandwest.com*

Marina services:

Moorage–summer transient available at times. **Power** 15 amp.
Launch ramp. Water, ice, tackle shop, **laundry, showers, toilets**. Chandlery. Accommodations (resort), RV park. Fishing gear and charters.
Nearby: Fuel: gas, diesel, stove oil. Restaurant, pub, bakery and other stores.

Ucluelet Public Docks

Boat Harbour– (inner harbour)
Box 910, Hemlock Street,
Ucluelet, BC. V0R 3A0.
(250) 726-4241 Fax (250) 726-7335.
Charts 3679, 3681
Fisheries & Oceans dock
Charts 3646, 3671, 3603
Manager • Float length 820 metres • Breakwater • Garbage/oil disposal • Derrick • Water • Lights • Power • Phone • Toilets • Showers • Near general stores, restaurants, Post Office, shops, laundry, accommodations, pool.

Other public docks (outer hbr)
Otter Street–
Fisheries and Oceans dock
Manager • Float length 150 m • Garbage disposal • Lights • Phone.
Itatsoo–Fisheries and Oceans dock
Manager • Float length 64 m • Wharf 36 m • Aircraft dock •
Main Street–Transport Canada dock
Manager • Float length 36 m • Wharf 42 m • Aircraft dock • Garbage disposal • Derrick • Lights • Power.
East–Fisheries and Oceans dock
Float length 49 m.

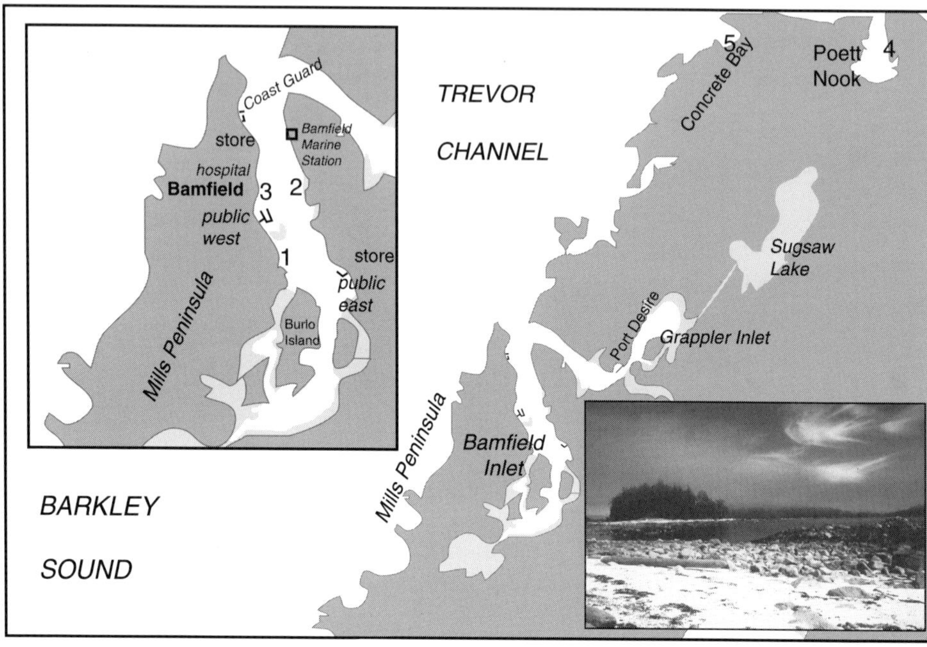

Bamfield

1. McKay Bay Lodge

Brian and Cheryl McKay
P.O. Box 116, Bamfield, BC.
V0R 1B0. Phone (250) 728-3323
Fax 728-3255
Chart 3646, 3671, 3602
Moorage–summer transient available. Fishing gear and charters. Fast-freeze. Accommodations (resort), restaurant, pub, bakery and other stores.

2. Bamfield Kingfisher Marina

Peter and Carol Brown
P.O. Box 38, Ucluelet, BC. V0R 1B0
Phone (250) 728-3228
Charts 3646, 3671, 3602
Fuel: diesel, gas. **Water.** Ice. Limited visitor moorage.

Bamfield Public Docks

East–Transport Canada dock
Charts 3646, 3671, 3602
Manager • Float length 128 metres • Wharf 18 m • Breakwater • Garbage disposal • Derrick • Lights • Phone • Sheds. Near general store.
Port Désiré–
Fisheries & Oceans dock
Manager • Float length 22 m • Launch ramp.
West–Fisheries and Oceans dock
Manager • Float length 154 m • Garbage
West–Transport Canada dock
Manager • Float length 24 m • Wharf 18 m • Garbage disposal • Derrick.

Barkley Sound is known for its fishing and scuba diving. The town of Bamfield is a good base for reaching some of the beautiful waters and beaches strewn throughout the Sound, especially in the islands of the nearby Broken Group. Photo above: Turret Island.

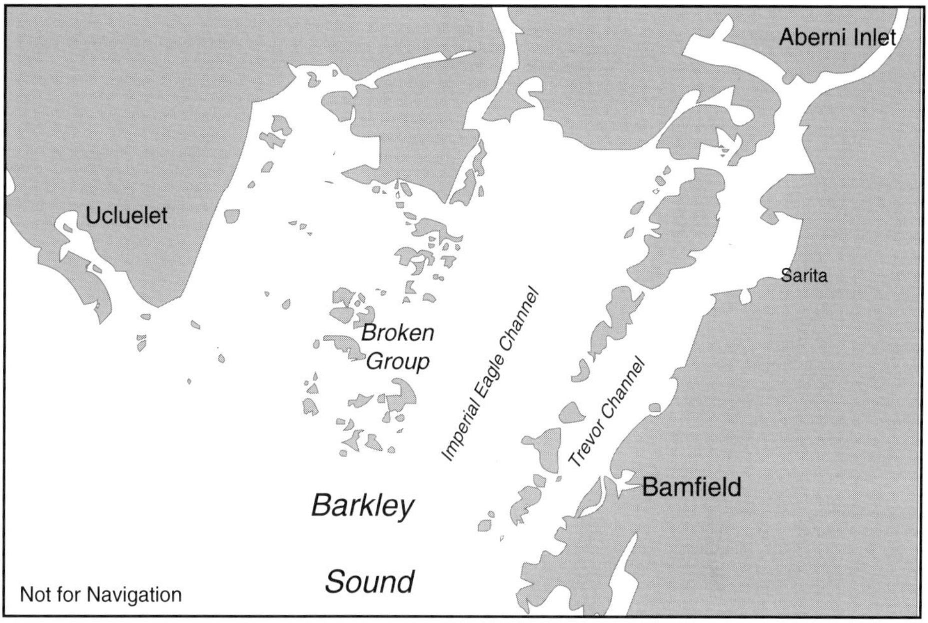

3. Bamfield Lodge and Cottages

Barry Otterson
P.O. Box 23, Bamfield,
BC. V0R 1B0 VHF 06
Phone (250) 728-3419 Fax 728-3417
Chart Charts 3646, 3671, 3602.
Website: *www.alberni.net/bamfieldlodge/*
Email: *bamlodge@cedar.alberni.net*
Moorage–summer transient when available. **Power**. **Water.** Ice. Fish freezers. Cappuccino bar.
Facilities–Accommodations (resort), cottages, charters, tours, marine field trips, kayaking, whale watching.

4. Poett Nook Marina

Stan and Flo Salmon
2178 Cameron Drive, Port Alberni,
BC. V9Y 1B2. Charts 3671, 3688
Phone (250) 724-8525
Marina: Moorage. Boats to 26 feet.
Fuel: gas.
Washrooms, laundry, showers.

5. Barkley Sound Resort

Craig Filipchuk
5425 Argyle Street, Port Alberni,
BC. VV9Y 1T6
Phone (250) 723-7462 Fax 723-3422
Chart 3671, 3668, 3602
Moorage, Accommodations, meals.
Power. **Water. Showers.**
Comfortable fishing boats are available 22–43 feet for charter.

Nearby

Accommodations (resort), restaurant, pub, RV park, bakery and other stores.
In Concrete Bay near Sarita and Poett Nook.

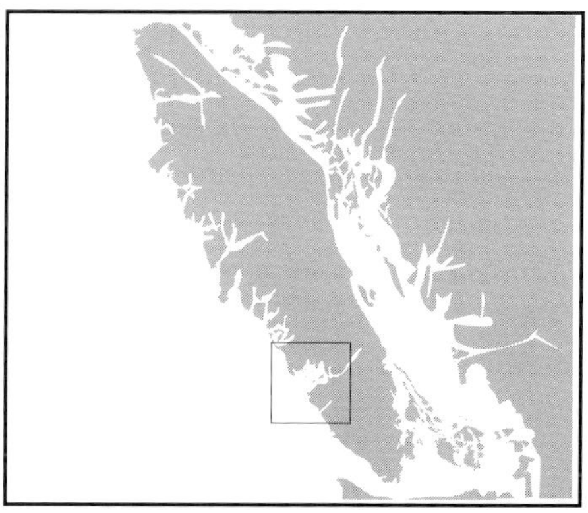

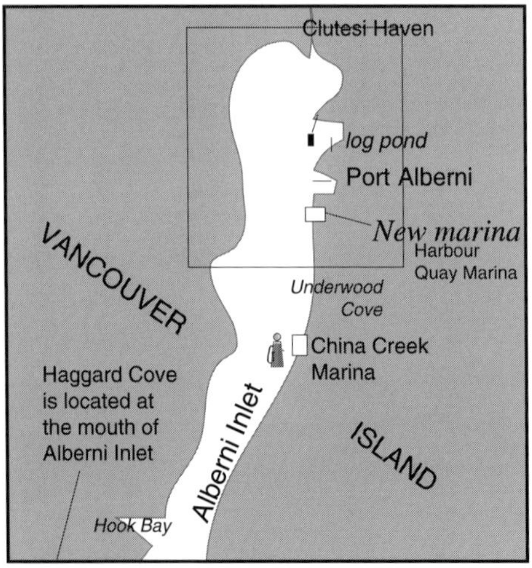

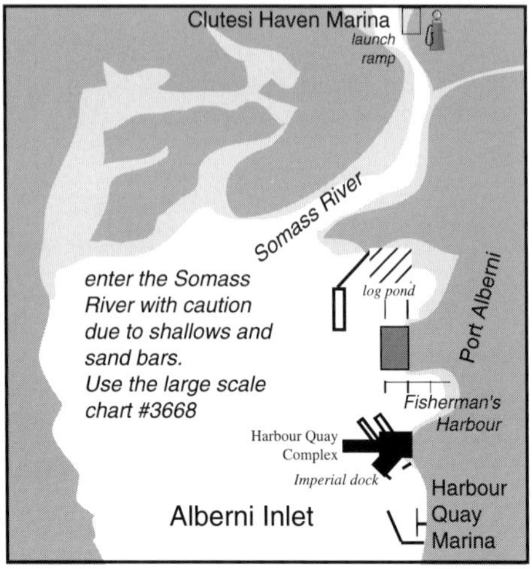

Port Alberni

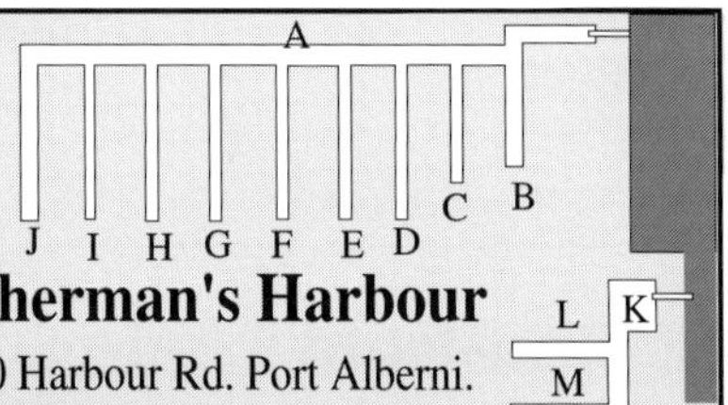

Fisherman's Harbour

3140 Harbour Rd. Port Alberni.
Fisheries and Oceans dock
Chart 3688
Managed by Port Alberni Harbour Commission–Bruce Kramer (wharfinger) • Float length 825 metres • Breakwater– 70 m wharf • Garbage disposal • Derrick • Water • Lights • Power • Phone. (250) 723-2533

Clutesi Haven

Tom McMillan
5104 River Road, Port Alberni,
BC. V9Y 6Z2 VHF 68
Phone (250) 724-6837 Fax 723-1114
Chart 3668 (up Somass River)
Moorage–summer transient available at times. **Water. Power** 30 amp.
Fuel: gas. Sani-dump. Phone.
Launch ramp. Ice, **laundry, showers, toilets**. Chandlery. Fishing gear and charters.
Nearby: Port Alberni: Grocery, hardware, propane, marine supplies, restaurants, bakery, liquor store and other facilities.

Haggard Cove Resort

Ron Clark
P.O. Box 396, Port Alberni,
BC. V9Y 7M9
Located near entrance to Alberni Canal.
Phone (250) 723-8457 Fax 723-5657
Chart 3668, VHF 27 (MV *Our Way*)
Moorage–Occasional summer transient available June to September.
Fishing Lodge. Fishing gear and charters.
Water. Accommodations, meals.

China Creek Marina

Roger Francouer
P.O. Box 575, Port Alberni,
BC. V9Y 7M9 VHF 16, 18A
Phone (250) 723-9812 Fax 723-9842
Chart 3668
Moorage–summer transient moorage if space available.
Fuel: diesel, gas. Water.
Power: 30 amps.

Customer Services:

Laundry. Showers. Washrooms.
Launch ramp adjacent. Ministore–fishing gear and supplies, cafe.
Campsite, sani-dump.

Harbour Quay Marina

Brian Stansbury

27590 Harbour Rd

Port Alberni V9Y 7X2, BC.

(250) 723-5312   Fax (250) 723-1114

cellular (250) 720-6256. **Chart 3688**

New harbour–breakwater and retaining shoreline have been completed with primary docks in place. The second phase plans are in place for 135 resident slips with power, water and all facilities. Visitor moorage will be available. Present moorage is available inside the breakwater or at the nearby Fishermen's Harbour. The local marine historical society and the Port Alberni Museum plan to mount the former Chrome Island lighthouse on top of a building going up on the pier.

Visit the quay at Fisherman's Harbour for restaurants, gift and craft shops. The Boathouse chandlery and marine service are nearby.

Inset above, left: The lauch ramp on the Somass River at Clutesi Haven in Port Alberni extends in the river (above). Fuel is available at China Creek Marina, below, and at Clutesi haven (bottom).

Inset above, right: At the Harbour Quay complex in Port Alberni.

Port Alberni Harbour Authority

Port Renfrew

Port Renfrew Marina

Gordon River Rd, Port Renfrew, BC. V0R 1B0
Phone (250) 478-3674 or 474-2643
Chart 3606 VHF 06

Enter up river on instructions from marina. Larger boats require high tide.

Bait and fishing licences available. Propane. Campground adjacent. **Launch ramp**. RV park.

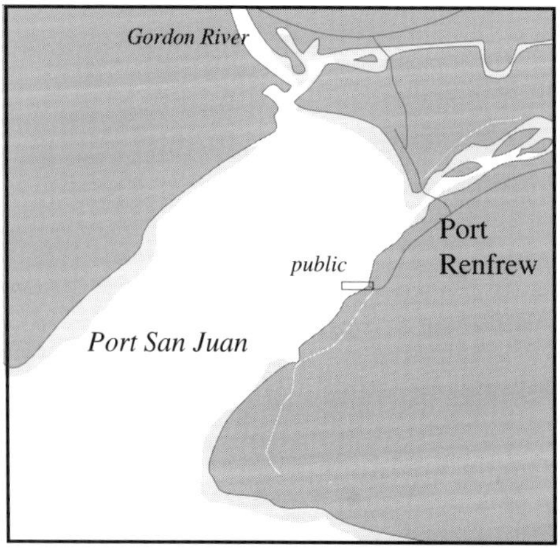

Port Renfrew Public Docks

East–Transport Canada dock
Charts
Manager • Float length • Wharf 33 m (Breakwater) • Derrick • Lights • Phone ashore • Pub. Restaurant nearby.

Above: The launch ramp at Port Renfrew is located on the Gordon River.
Opposite: The public docks at Sooke are located on the west shore of the harbour just after entering past Whiffen Spit. Exercise great caution entering Sooke, using the channel markers to avoid hazards at the entrance.

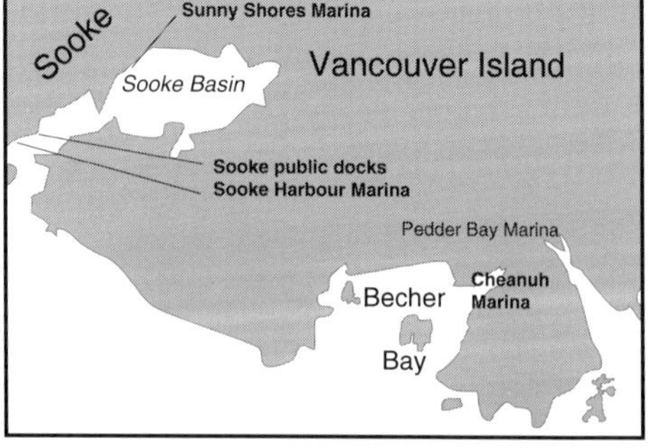

Sooke Harbour Marina

Sooke

Sooke Harbour Marina

Larry and Darlene Doucette
6971 West Coast Road,
Sooke, BC. V0S 1N0
Phone (250) 642-3236
Chart 3641 VHF 16 &68
Marina: Power 15 amp. **Water. Ice. Laundry. Showers. Washrooms.**
Moorage–limited summer transient. Phone, Camping. Launch Ramp. Fuel–2 km by road, also stores and restaurants.
Beware of sand bank out in channel. Follow markers and use marine chart.

Sunny Shores Marina

Andrew Planeta
5621 Sooke Road, Sooke,
BC. V0S 1N0 Ph: 1-888-805-3932
Phone (250) 642-5731 Fax 642-5737
email: *sunnyshores@sookenet.com*
Chart 3606 *(Enter from Eliza Point favouring the eastern shore)*
Fuel gas, diesel, oil**. Power. Water. Launch Ramp.** Fishing licences.
Moorage–some summer transient slips.
Showers. Washrooms. Laundry.
Mechanical service. Ministore–groceries. Hotel and camping.
Nearby restaurants and bus service to Sooke and Victoria.

Sooke Public Docks

Fisheries and Oceans dock
Chart 3641
Manager • Float length 243 metres • Wharf 34 m • Garbage disposal • Derrick • Water • Lights • Power • Easy walk up into town for all shops and services.

Sunny Shores Marina

Cheanuh Marina

Stan Chipps
Box 4, 4901 East Sooke Road,
RR#1 Sooke, BC. V0S 1N0
Phone (250) 478-4880 Fax 478-3585
Chart 3641 Located in Becher Bay
(Enter with caution)

Fuel gas only. **Moorage**–some summer transient available–call for reservations. Boat rentals. Fishing and camping.

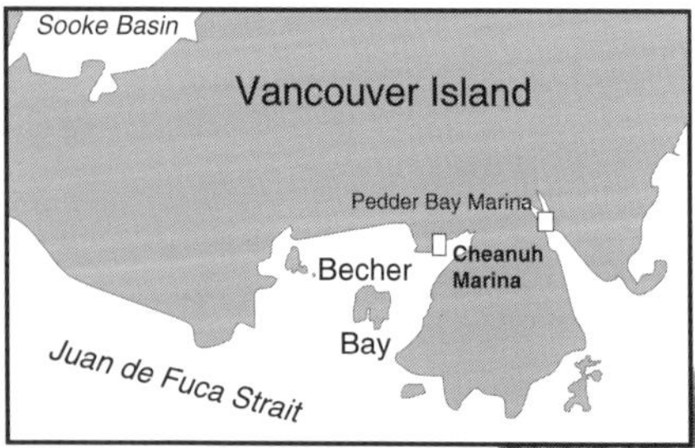

Pedder Bay Marina

Sean Moore
925 Pedder Bay Drive, No 12
Metchosin, Victoria, BC. V9Z 4H1
Phone (250) 478-1771 Fax (250) 478-2695
Fax (250) 728-3255
Chart 3410 VHF 68

Moorage–some summer transient. Fishing licences. Phone, washrooms, showers, laundry. Rental boats and fishing guides available. Camping, mini chandlery. **Launch Ramp.**

Beware of the reef at entrance to marina.

Top:A view of Cheanuh Marina in Becher Bay. Right: Pedder Bay Marina

Index

Bibliography

99 Dives. San Juan Islands, Gulf Islands and Vancouver Island. Betty Pratt-Johnson. Heritage House, Surrey, BC. 1994.
101 Dives. From the mainland of Washington and British Columbia. Betty Pratt-Johnson. Heritage House, Surrey, BC. 1994.
A Guide to the Western Seashore. Rick M. Harbo. Hancock House, Surrey, BC. 1988.
BC Cruising Guide Series–Desolation Sound, Gulf Islands, Sunshine Coast. Whitecap Books, North Vancouver, BC. Bill Wolferstan.
Charlies Charts North to Alaska. Charles E. Wood. Margo Wood. Polymath Energy Consultants Ltd. Surrey B.C. 1986.
Canadian Tide and Current Tables. Pacific Coast all volumes. Ottawa-Department of Fisheries and Oceans. Annual.
Cruising Beyond Desolation Sound. John Chappel/ Naikoon Marine, Surrey, BC. 1979.
Docks and Destinations. Guide to marinas in British Columbia. Peter Vassilopoulos. Seagraphic Publications. 1995.
Exploring Vancouver Island's West Coast. A Cruising Guide. Don Douglass. Fine Edge Productions. 1994.
Exploring the Inside Passage to Alaska. A Cruising Guide. Don Douglas. Fine Edge Productions. Bishop, California.1995.
Exploring the South Coast of British Columbia. A Cruising Guide. Don Douglas. Fine Edge Productions. Bishop, California. 1996.
Exploring the Gulf Islands and Desolation Sound to Port Hardy and Blunden Harbour. A Cruising Guide. Don Douglas. Fine Edge Productions. Bishop, California. 1996.
Marine Parks of British Columbia. An Explorer's Guide. Peter Chettleburgh. Special Interest Publications. Vancouver. BC. 1985.
Marine Weather Hazards Manual. A guide to local forecasts and conditions. Vancouver. Environment Canada. 1990.
North to Alaska. Hugo Anderson. Anderson Publishing Co. 1993
Northwest Boat Travel. Anderson Publishing Company. Anacortes. Annual.
Oceanography of the British Columbia Coast. Richard E. Thomson. Ottawa. Department of Fisheries and Aquatic Sciences. 1981.
Sailing Directions. British Columbia Coast (North Portion). Ottawa. Department of Fisheries and Oceans.
Sailing Directions. British Columbia Coast (South Portion). Ottawa. Department of Fisheries and Oceans.
Sea Kayaking Canada's West Coast. John Ince and Hedi Kottner. Raxas Books. Vancouver, BC. 1996.
The San Juan Islands. Afoot and Afloat. Marge and Ted Mueller. The Mountaineers. Seattle. 1988.
Waggoner. Robert Hale. Robert Hale Publishing. Seattle. Annual. *This is an excellent cruising guide with up-to-date information about marinas and other facilities. It includes planning and piloting information.* Ph (800) 733-5330.
Weatherly Waypoint Guides for GPS and Loran Navigation. Robert Hale. Robert Hale Publishing. Volumes 1–3: Puget Sound, San Juan Islands, Strait of Juan de Fuca. Gulf of Georgia, including Gulf Islands, Jervis Inlet. Desolation Sound to Port Hardy.
West Coast of Vancouver Island. Don Watmough. Whitecap Books, North Vancouver, BC.

My wife Carla and I have cruised the area this guide covers over the past twenty-five years. We have visited all areas described in the book and have stopped at and anchored in most anchorages included in the foregoing pages. Mariners who adventure beyond the known routes and popular areas will enjoy discovering for themselves others I have omitted. There are numerous books on cruising the coast and these along with your charts and reference books should enable you to extend your cruising range substantially and safely. Happy boating–Peter Vassilopoulos.

99 Dives, 101 Dives
Suba diving guides to west coast dive sites–by Betty Pratt Johnson
$28.95 & $29.95

Also available from Seagraphic Publications:

Antiques Afloat–*by Peter Vassilopoulos* *$19.95* *(soft cover) plus $2.50 S&H.*

This book, published in 1981, is a coffee table book containing a colourful collection of boats that plied the waters of British Columbia in the early part of the 20th Century. The boats included in the book were representative of an era in which wooden craftsmanship was at a peak and large, opulent vessels were a reflection of the success of their owners. Among those included were the vessels owned by Mr and Mrs Boeing, a former lieutenant governor, the Rogers family and others including some notorious rum runners. Some hard cover copies are still available direct or at some marine stores. $29.95. *(Both editions offered while limited stocks last.)*

Anchorages and Marine Parks–*by Peter Vassilopoulos* *$19.95* *(plus $2.50 S&H)*

This is a companion guide to **Docks and Destinations**. It is a *user friendly guide* to places to find sheltered anchorage overnight and to facilities and features of marine parks. The area it covers, in a south to north progression, includes the San Juan Islands and all of BC coastal waters including the west coast of Vancouver Island. The contents includes availability at marine parks by way of descriptive icons showing water, showers, toilets, picnic tables, launch ramps, camping, hiking trails and much more. It contains numerous drawings depicting anchorage layouts and their relative location on the coast and is illustrated with hundreds of aerial and other photographs. It is a best selling publication and is also available from all leading marine stores, marinas and book stores.

101 Dives $29.95 and 99 Dives $28.95 –Betty Pratt-Johnson *(plus $2.50 S&H ea)*

Published by heritage House. Two excellent guides to diving the waters of British Columbia and Washington State.

DIVER Magazine–*Peter Vassilopoulos. Seagraphic Publication $3.50 per copy*

This Canadian magazine is published nine times a year. It contains a wealth of information about marine life, shipwrecks, underwater photography, diving medicine, personal profiles, dive travel destinations and much more. It is a national publication and covers BC extensively as well as other parts of Canada where diving is popular. It features regular dive destinations in tropical waters and provides lots of high quality colour illustrations. It is available at dive stores and selected news stands. Or by subscription or single copy from the publisher.

Subscribe at a special trial rate of $15.00 for one year–9 times a year. (includes taxes).

Order any of these books or magazines by sending payment to the publisher. Please include $2.50 for shipping and handling for books. No extra for magazine subscriptions.

Contact Seagraphic Publications
Box 1312 Delta B.C. V4M 3Y8 (604) 274-4333
or Box 984, Point Roberts, WA. 98281-0984

United States orders in US funds amounts as shown. Canadians please pay in Canadian funds amounts shown.

Docks and Destinations

Some comments about Docks & Destinations (and Anchorages & Marine Parks)

We often refer to our copy of **Docks and Destinations** while out on the water. We ALWAYS refer to your book when we are working in our store. What a great book!
–Dan and Leah Lee, Thrifty Foods, Salt Spring Island.

Peter Vassilopoulos' nautical guidebooks are among the top ten trade bestsellers in our store this summer (1998). That's the verdict from the public!
–Merv Adey, Tanners Books, Sidney, B.C.

Our best and most informative cruising guide. Always our first recommendation to our customers looking for a cruising guide on the coast. (We use it on our own boat.)
–Brad Mah, Nikka Industries, Vancouver/Steveston.

Mariners have commented on the clarity of your book. It is well designed—succinct yet containing enough information to enable the boater to make wise cruising decisions.
–Ann Taylor, Greenway Sound.

The book has helped us find new destinations we had never heard of before. Thank you for the best boating guide on the coast.
–Jeanne and Henry Karcz, Richmond, BC.

This guide is updated when reprinted. Major updates and changes will be made periodically when new editions are published. Please write to me if you have any information or suggestions for inclusion in future editions. Your comments will be welcome.
–Peter Vassilopoulos.

The author's Monaro 27, used also for work with his Canadian national scuba diving publication, DIVER Magazine.

Canada: P.O. Box 1312 Delta, Stn A., British Columbia, Canada. V4M 3Y8
USA: P.O. Box 984, Point Roberts, WA. 98281-0984
Phone (604) 274-4333 or (604) 943-4198 Fax (604) 274-4366
email: *divermag@axionet.com* (Please contact us for updating information.)